FM 25-5

US WAR DEPARTMENT BASIC FIELD MANUAL

ANIMAL TRANSPORT

1939

CIVILIAN REFERENCE EDITION

UNABRIDGED HISTORIC CALVARY MANUAL ON HORSES FOR RIDING AND WORK,
INCLUDING HORSE CARE, RIDING, WAGONS, PACKS,
AND OTHER TRADITIONAL EQUESTRIAN TECHNIQUES

U.S. WAR DEPARTMENT

Doublebit Press

*Original content under the public domain; unrestricted for civilian distribution.
Originally published in 1939 by the U.S. War Department.*

A Part of the Military Outdoors Skills Series

*Doublebit Press Civilian Reference Edition ISBNs
Hardcover: 978-1643892023
Paperback: 978-1643892030*

The Military Outdoors Skills Series

Historic Field Manuals and Military Guides on Outdoors Skills and Travel

Military manuals contain essential knowledge about outdoors life, thriving while in the field, and self-sufficiency. Unfortunately, many great military books, field manuals, and technical guides over the years have become less available and harder to find. These have either been rescinded by the armed forces or are otherwise out of print due to their age. This does not mean that these manuals are worthless or "out of date" – in fact, the opposite is true! It is true that the US Military frequently updates its manuals as its protocols frequently change based on the current times and combat situations that our armed services face. However, the knowledge about the outdoors over the entire history of military publications is timeless!

By publishing the **Military Outdoors Skills Series**, it is our goal at Doublebit Press to do what we can to preserve and share valuable military works that hold timeless history about outdoors life, navigation, and survival. These books include official unrestricted texts such as army field manuals (the FM series), technical manuals (the TM series), and other military books from the Air Force, Navy, and texts from before 1900. Through remastered reprint editions of military handbooks and field manuals, outdoors enthusiasts, bushcrafters, hunters, scouts, campers, survivalists, nature lore experts, and military historians can preserve the time-tested knowledge and history about the skills and institutional procedures that were learned through hard lessons and training by the U.S. Military and our expert soldiers.

This book is an important contribution to self-sufficiency literature and has important historical and collector value toward preserving these traditions.

Although the editors at Doublebit Press are thrilled to have comfortable experiences in the woods and love our high-tech and light-weight equipment, we are also realizing that the basic skills taught by the old experts are more essential than ever as our culture becomes more and more hooked on digital technology. We don't want to risk forgetting the important steps, skills, or building blocks involved with thriving in the outdoors. This Civilian Reference Edition historic reprint is a part of a collection of military handbooks and field manuals that are essential contributions to the American self-sufficiency tradition despite originating with the military. In the most basic sense, these books are the collection of experiences by the great experts of outdoors skills and self-sufficiency life: our countless expert soldiers who learned to thrive in the backwoods, deserts, extreme cold environments, and jungles of the world.

With technology playing a major role in everyday life, sometimes we need to take a step back in time to find those basic building blocks used for gaining mastery – the things that we have luckily not completely lost and has been recorded in books over the last two centuries. These history hasn't been forgotten, it's just been shelved. *It's time to unshelve this history once again and reclaim the lost knowledge of self-sufficiency.*

Based on this commitment to preserving our self-sufficiency heritage, we have taken great pride in publishing this book as a complete, unabridged and original work. We hope it is worthy of both study and collection by outdoors folk in the modern era of self-sufficiency and traditional skills life.

Unlike many other photocopy reproductions of classic books that are common on the market, this historic Civilian Reference Edition does not simply place poor photography of old texts on our pages and use error-prone optical scanning or computer-generated text. We want our work to speak for itself, and reflect the quality demanded by our customers who spend their hard-earned money. With this in mind, each Historic Edition book that has been chosen for publication is carefully remastered from original print books, *with the Doublebit Civilian Reference Edition printed and laid out in the exact way that it was presented at its original publication*. We provide a beautiful, memorable experience that is as true to the original text as best as possible, but with the aid of modern technology to make as beautiful a reading experience as possible for books that are typically over a century old. Military historians and outdoors enthusiasts alike are sure to appreciate the care to preserve this work!

Because of its age and because it is presented in its original form, the book may contain misspellings, inking errors, and other print blemishes that were common for the age. However, these are exactly the things that we feel give the book its character, which we preserved in this Historic Edition. During digitization, we ensured that each illustration in the text was clean and sharp with the least amount of loss from being copied and digitized as possible. Full-page plate illustrations are presented as they were found, often including the extra blank page that was often behind a plate. For the covers, we use the original cover design to give the book its original feel. We are sure you'll appreciate the fine touches and attention to detail that your historic Civilian Reference Edition has to offer.

For outdoors, self-sufficiency, and military history enthusiasts who demand the best from their equipment, the Doublebit Press Civilian Reference Edition reprint of this military manual was made with you in mind. Both important and minor details have equally both been accounted for by our publishing staff, down to the cover, font, layout, and images. It is the goal of Doublebit Civilian Reference Edition series to preserve outdoors and self-sufficiency heritage, but also be cherished as collectible pieces, worthy of collection in any library and that can be passed to future generations.

FM 25-5

BASIC FIELD MANUAL

ANIMAL TRANSPORT

Prepared under direction of the
Chief of Cavalry

UNITED STATES
GOVERNMENT PRINTING OFFICE
WASHINGTON : 1939

For sale by the Superintendent of Documents, Washington, D. C., Price 25 cents

WAR DEPARTMENT,
WASHINGTON, *June 15, 1939.*

FM 25-5, Basic Field Manual, Animal Transport, is published for the information and guidance of all concerned.

[A. G. 062.11 (3-1-38).]

BY ORDER OF THE SECRETARY OF WAR:

MALIN CRAIG,
Chief of Staff.

OFFICIAL:

E. S. ADAMS,
Major General,
The Adjutant General.

TABLE OF CONTENTS

BASIC FIELD MANUAL

ANIMAL TRANSPORT

(The matter contained herein supersedes TR 50-45, February 21, 1922 (including C1, January 2, 1926); TR 360-5, November 5, 1925 (including C1, January 3, 1928, C2, January 2, 1929, and C3, February 15, 1933); TR 360-10, March 28, 1922 (including C1, January 2, 1926); TR 360-20, January 11, 1926; TR 360-25, November 1, 1926 (including C1, January 2, 1930); TR 380-10, November 5, 1925 (including C1, January 3, 1927, C2, January 2, 1929, and C3, January 2, 1931); TR 380-15, January 20, 1936; TR 400-50, January 4, 1926 (including C1, January 2, 1929); TR 400-60, January 20, 1926; and Chapters 1 to 5, inclusive, Basic Field Manual, Volume V (tentative), 1929.)

CHAPTER 1

HORSEMANSHIP

SECTION I

GENERAL PRINCIPLES

1. Purpose of instruction.—The purpose of individual instruction mounted without arms is to teach so much of the principles and methods of horsemanship as is necessary to permit the soldier to manage his horse in ranks, to ride alone on such missions as he may be called upon to perform as a mounted soldier, and later to use his weapons in mounted combat.

2. Instructors.—The instructor should be a skilled and experienced horseman, properly mounted. His manner should always be quiet and deliberate. In his instruction he should present his ideas in logical order, and supplement the original explanation of a movement by executing it himself. A sufficient variety of instruction must be given in each period to retain interest. In order to avoid soreness, no one exercise should be extended over an unduly long period.

3. Recruits.—The confidence of the recruit in his ability to learn to ride must be carefully nurtured. Fear must be banished from his mind before appreciable progress can be made. The average recruit is not in excellent physical condition and as his physique can be developed only gradually, frequent rest periods must be given.

4. Nature of horses.—A knowledge of the nature and gaits of horses is of great assistance to men in developing their confidence. It should be explained to all recruits that horses are timid animals that react to hunger and fear, and their natural defense is in flight. Contrary to popular belief, horses do not think for themselves but are creatures of habit. Habits are developed during their training by taking advantage of their hunger and fear. Horses are naturally afraid of being hurt and of strange sights and sounds. Much of their early training is accomplished through the infliction of a small amount of pain by use of the legs or light taps with a whip, or pain on the bars of the mouth resulting from use of the reins. As their training progresses, they respond to the signals given by the legs or reins in anticipation of being hurt if they disobey. Finally the trained horse obeys through habit. However, if startled or roughly handled, fear may cause them to run, buck, or kick. All recruits should be cautioned to avoid sudden movements on or around horses.

5. Speed of various gaits.—In the military service it is essential for the rider to maintain certain rates of speed when at the walk, trot, or gallop.

a. The rate of the walk is 4 miles per hour, or 117 yards per minute. At the commands 1. ***Forward,*** 2. ***MARCH,*** the walk is taken and maintained.

b. The rate of the trot is 9 miles per hour, or 264 yards per minute. At the commands 1. ***Trot,*** 2. ***MARCH,*** the trot is taken and maintained. For the purpose of individual instruction in equitation, the rate of the trot may, at command, be reduced to 6 miles per hour. The commands are: 1. ***Slow trot,*** 2. ***MARCH.*** When at the slow trot the rider should sit down in the saddle; when at the trot he should rise to the trot.

c. The gallop is at the rate of 12 miles per hour, or 352 yards per minute. At the commands 1. ***Gallop,*** 2. ***MARCH,*** the gallop is taken and maintained.

6. Programs.—Local conditions influence greatly the type of program which should be followed with any particular group of recruits. Time available is often a limiting factor, but it is highly desirable that not less than 8 weeks be devoted to this instruction. If less time is available, the instructor must judge how the instruction should be limited. In any case instruction must progress in a logical sequence.

7. Schedules.—Daily or weekly schedules must be varied to maintain interest of both men and animals. A certain amount of time each day must be devoted to review of previous work.

8. Military seat.—"Seat" is the term used to describe the posture or way of sitting on a horse. A good seat enables the rider to maintain his balance whatever may be the action of his horse. Such a seat depends upon practice, suppleness, and muscular control. The rear limit of the military seat is at the points of the buttocks. With the points of the buttocks resting lightly in the saddle, its throat deep in the rider's crotch, the naturally erect upper body of the rider is so inclined from the hips as to maintain balance, carry the weight into the lower thighs, and force them down into their proper positions. The fleshy parts of the buttocks are to the rear and in no case form part of the seat. The correct position of the rider's back and loins is aided materially at all gaits by keeping the head, chin, and chest up.

9. Position mounted.—*a. At the halt.*—(1) The rider sits squarely in the middle of the saddle, upper body erect, his weight distributed from the points of his buttocks forward upon his crotch, inner thighs, knees, and stirrups.

(2) Head and chin are up, neck erect without stiffness. Eyes look to the front.

(3) The back is without stiffness in a position identical with that of the dismounted soldier at attention. The center of gravity of the upper body is always kept in front of the buttocks. The chest is lifted, and shoulders square without stiffness. Arms are free and relaxed, elbows hanging naturally by the sides.

(4) Hip joints are relaxed to enable the rider to maintain balance.

(5) Thighs extend downward and forward, their inner sides resting without constraint on the saddle.

(6) The knees, flexed and relaxed, are forced down as low as the adjustment of the stirrups will permit. They rest with their inner sides in continuous contact with the saddle.

(7) (*a*) The lower legs extend downward and slightly backward with the calves in light elastic contact with the horse.

(*b*) Stirrup leathers are approximately vertical. The lengths of stirrup leathers are correctly adjusted when the treads of the stirrups are nearly opposite the lower edges of the ankle bones of the rider, seated as prescribed, with feet out of stirrups and legs hanging naturally.

(*c*) When using the stirrup, the broadest part of the foot rests upon the stirrup tread. Heels are well down, toes turn out naturally, and ankles are flexed and relaxed. Without stirrups the legs hang naturally, knees and ankles relaxed, toes lower than heels, and feet turned out in a natural position.

(8) The reins may be held in either or both hands, fingers closed without stiffness, knuckles about 30° from the vertical. The wrist is straight and supple, and the forearm, wrist, hand, and rein form one straight line from point of elbow to horse's mouth, the elbow being slightly in advance of the point of the hip. With the reins in one hand, the free arm hangs naturally, hand in rear of the thigh. When both hands are used, they are well separated and normally held above the withers.

b. In motion.—Upper body at various gaits:

(1) At the walk the upper body is inclined forward very slightly from its position at the halt. The upper body has the same erect appearance as that of the soldier dismounted at attention.

(2) At the slow trot, with or without stirrups, the upper body remains erect without stiffness, and with sufficient forward inclination to keep its center of gravity over its base of support. The forward inclination is only slightly greater than at the walk, and the rider has the appearance of sitting erect.

(3) At the posting trot the rider's body moves upward and forward and downward and backward in cadence with and

actuated by the beats of the gait. Sufficient forward inclination is taken to be in balance. The upper body, pivoting at the hip joints, maintains its posture unchanged without sinking rearward at the loins, as the rider comes into the saddle at each alternate beat of the trot. The chin is raised so that the plane of the face remains vertical. The rider sinks into the saddle very lightly on the upper thighs and crotch, points of buttocks barely touching the saddle at each beat.

(4) At the gallop the upper body is inclined forward slightly more than as described for the walk or slow trot. At the extended gallop or charge, the upper body is inclined forward with the buttocks entirely out of the saddle, and the weight on the lower thighs, knees, and stirrups.

(5) When changing gaits or rates, the degree of forward inclination of the upper body is dependent upon the suddenness

FIGURE 1.—Position of the soldier, mounted.

of the change. When increasing the gaits, the inclination is sufficient to prevent the center of gravity of the upper body from falling in rear of the base of support. In decreasing the gaits, the upper body becomes more erect so that the center of gravity may remain in rear of the knees. In either case the change of position is reduced to the minimum required for remaining in balance.

10. Aids in controlling horse.—The term "aids" implies the separate or collective application of the voice, weight, reins, legs, or legs and spurs in controlling the horse.

a. Voice.—The voice is a simple and effective aid. Its effect lies in intonation rather than words or phrases spoken. Low, soothing tones have a tendency to quiet the horse, whereas sharp, higher pitched tones encourage him to greater activity. Clucking is an effective method to encourage the horse to move forward or go faster.

b. Weight.—The proper use of the weight of the rider's upper body assists the horse in performing a desired movement. In increasing the gait, the upper body is inclined forward to allow the rider to maintain his balance. This allows the horse to move more easily to the front. Similarly, if the rider lessens the forward inclination of the upper body, it will assist the horse in slowing the gait.

c. Reins.—Increased tension of the reins directly to the rear causes the horse to decrease the pace, to change to a slower gait, to halt, or to back. If the horse is standing still, an action of the reins and legs prepares him to move. Relaxing the tension on the reins combined with closing the legs causes him to move forward or, if already moving, to increase the speed of the gait. Relaxing the reins means a relaxation of the fingers which lessens the pressure of the bits on the horse's mouth. The action of one or both reins toward one side causes the horse to change direction correspondingly.

(1) The *direct rein* is produced by fixing the hands and closing the fingers on the adjusted rein, thus increasing tension on the rein. Elbow and hand should not move to the rear. The two reins acting together as direct reins have the effect of slowing, stopping, or backing the horse.

(2) The *leading rein* (the right, for example) is produced by moving the right hand to the right without increasing ten-

sion on the reins, thus displacing the horse's head to the right and causing the horse to turn to the right. The rider should avoid pulling on the rein from front to rear, or opening the elbow away from the body.

(3) The *bearing rein* (the right, for example) is produced by moving the right hand to the left so that the rein bears against the neck without increased tension to the rear. The bearing rein draws the horse's muzzle to the right, forces the neck at the shoulders to the left, and loads the left shoulder with the greater part of the weight on the neck. The increase of weight breaks the equilibrium and the horse turns toward the left.

Caution: The hand should never be carried over the midline of the horse's neck.

d. Legs.—(1) The lower legs should be fixed in light elastic contact with the sides of the horse, free from all involuntary motion and definite in their application. The legs may act, resist, or yield. They act when their pressure increases to determine a movement; they resist when their pressure remains constant and is opposed to a displacement at the hindquarters; they yield when their pressure diminishes and allows the displacement.

(2) When the two legs act simultaneously slightly in rear of the girth, the effect, if at a halt, is to move the horse forward; if in march, to increase the speed or gait.

(3) When one leg acts alone in rear of the girth, it provokes forward movement while pushing the hindquarters to the opposite side.

(4) If increased pressure is insufficient, the legs should act by light taps of the calf, and should cease to act as soon as the horse obeys. The heels should be kept low; heels or spurs should not be used unless the horse fails to obey the legs.

e. Leg and spur.—The spur should not be used at all if the same result can be obtained by use of the calf of the leg. When necessary the spur should be used just in rear of the girth by rotating the ankle without depressing the toe or raising the heel, and with a series of successive touches rather than a constant pressure. These should cease as soon as the horse obeys.

f. Manner of applying the aids.—The action of the legs and reins should be intermittent, not continuous, and with an intensity only great enough to produce the desired effect. When the effect has been gained, the horse should immediately be rewarded by the rider maintaining a light feel on his mouth sufficient only to keep the horse in the proper position. If an action of the aids is continuous and prolonged, the horse becomes accustomed to the effect or has time to adjust himself so as to set up resistance.

g. Use of aids in combination.—An aid is seldom employed singly, but in combination with others. One aid does not need to act with as much force if assisted by another. In using the aids in combination, the rider must be careful that the effect of one aid is not opposed to that of another.

h. To keep hand steady.—The hand should be free of all involuntary and useless movement, wrist kept straight but relaxed, and fingers and elbows so employed as to allow the rider to maintain an even tension on the reins.

SECTION II

INSTRUCTION, MOUNTED

11. Fitting of horse equipment.—*a.* Proper fitting of bits and saddles should receive systematic, continuous attention. Habitual attention to the ordinary measures for the prevention of sore backs, and adjustments for avoidance of unnecessary irritation to the animal are indispensable to the efficiency of a mounted organization.

b. The following principles are of general application in saddle fitting:

(1) The withers must not be pinched nor pressed upon.

(2) The central line of the back must have no pressure upon it.

(3) The shoulder blades must have free and unhampered movement.

(4) The weight must be evenly distributed over a surface which extends from the rear of the shoulders to the last true rib.

c. The bits should be of proper width to fit the horse's mouth, the branches of the curb touching lightly the outer surfaces of the lips.

12. To fold the saddle blanket.—*a*. The blanket, after being well shaken out, should be folded into six thicknesses, as follows:

(1) Hold up by two corners, so that long edge of blanket is up and down.

(2) Double lengthwise, folded corner between thumb and forefinger of right hand, thumb pointing to left.

(3) Slip left hand down folded edge two-thirds its length and seize with thumb and second finger.

(4) Raise hands to height of shoulders, blanket extended between them.

(5) Bring hands together, double fold falling outward.

(6) Pass folded corner from right hand into left hand.

(7) Slip second finger of right hand between folds and grasp double-folded corners.

(8) Turn left (disengaged) corner in and grasp with thumb and forefinger of right hand, the second finger of right hand stretching and evening the folds.

(9) Grasp corners and shake blanket well in order to smooth the folds.

(10) Raise blanket and place upper edge between the chin and breast.

(11) Slip hands halfway down blanket, first two fingers outside, other fingers and thumb of each hand inside.

(12) Grasp blanket and let the part under the chin fall forward.

(13) Hold blanket up, arms extended, even the lower edges, then flip outside part over right arm.

(14) Hold blanket thus until ready to place it on the horse.

b. The description given in *a* above is the normal method of folding the saddle blanket. Departure from this method may be resorted to, however, in order to prevent undue wear on the bearing surface and along the folded edges of the blanket and, on occasion, for the relief or prevention of sore backs.

13. To stand to heel.—The horses being tied on the picket line or in their respective stalls in the stable, at the command ***Stand to heel,*** each man stands at attention 1 yard in rear of and facing his horse.

14. To put on blanket and surcingle.—The instructor commands: ***BLANKET.***

a. Approach horse on left side, with blanket folded and held as described in paragraph 12.

b. Place blanket well forward on horse's back by tossing the part of blanket over right arm to right side of the horse, still keeping hold of the middle points.

c. Slide blanket once or twice from front to rear to smooth the hair on the horse's back, being careful to raise blanket in bringing it forward.

d. Place blanket with forefinger of left hand on the withers and forefinger of right hand on the loin.

e. Pass buckle end of surcingle over middle of blanket.

f. Buckle surcingle on left side of the horse a little below the edge of blanket.

15. To saddle.—*a. McClellan saddle.*—(1) For instruction, the saddle may be placed 4 yards in rear of the horse, the stirrups crossed over the seat, right one uppermost; the cinch and cinch strap crossed above the stirrups, the strap uppermost; saddle blanket folded, horse side in, folded edge to the front, double fold to the right, across the seat of the saddle.

(2) The instructor commands: ***BLANKET.*** Approach horse on left side, with blanket folded and held as described in paragraph 12. Place blanket (less surcingle) as described in paragraph 14.

(3) The blanket having been placed on the horse, the instructor commands: ***SADDLE.***

(*a*) Grasp pommel of saddle with left hand and cantle with right.

(*b*) Approach horse on left side and place center of saddle on the middle of the horse's back, front end of side bar about four fingerbreadths behind the point of shoulder blade and about the width of a man's hand in rear of double-folded edge of blanket.

(*c*) Let down cinch strap and cinch.

(*d*) Pass to right side, adjust cinch and strap, and see that blanket is smooth.

(*e*) Return to left side and raise blanket slightly under pommel arch so that the withers will not be compressed.

(*f*) Take cinch strap in right hand and reach under the horse and seize cinch ring with left hand.

(*g*) Pass end of strap through ring from underneath (from inside to outside), then up and through upper ring from the outside; if necessary, make another fold in the same manner. Do not get cinch too tight. When adjusted, the flat open hand should fit snugly between the cinch and the horse.

(4) Inspect to see that the withers are not pressed or pinched.

b. Modified McClellan saddle.—When the modified McClellan saddle is used, the instructions given in *a* above govern except as indicated below:

(1) After saddle has been placed on the horse's neck, pass to right side and see that blanket is smooth, and let down the girth.

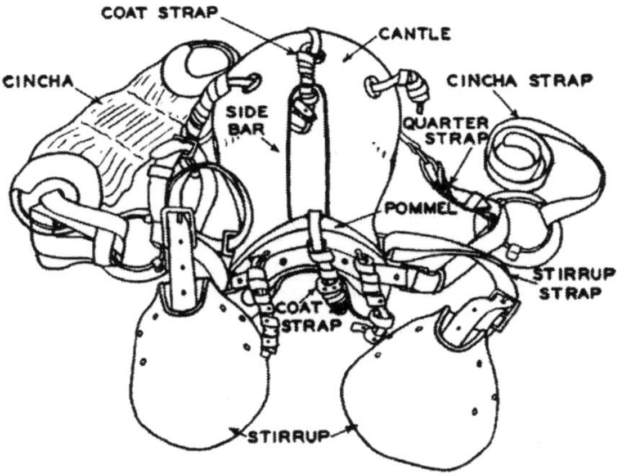

FIGURE 2.—Parts of saddle, McClellan, M1904, cavalry.

(2) Return to left side and raise blanket slightly under pommel arch. Reach under horse and seize end of girth with left hand. Raise saddle skirt with right hand. Insert front girth strap in front buckle of girth and draw up snugly. Repeat the same operation with rear strap and rear buckle.

16. To adjust length of stirrup straps.—To approximate the correct length of stirrup straps, before mounting, they are adjusted so that the length of stirrup strap, including stirrup, is about one inch less than the length of the arm, fingers extended.

17. To unsaddle.—*a. McClellan saddle.*—The instructor commands: ***UNSADDLE.***

(1) Cross left stirrup over saddle and loosen cinch strap and let down cinch.

(2) Pass to right side and cross right stirrup, then cinch.

(3) Pass to left side and cross cinch strap over saddle.

(4) Grasp pommel with left hand, cantle with right hand and, removing saddle toward croup, place it in rear of the horse, pommel to front.

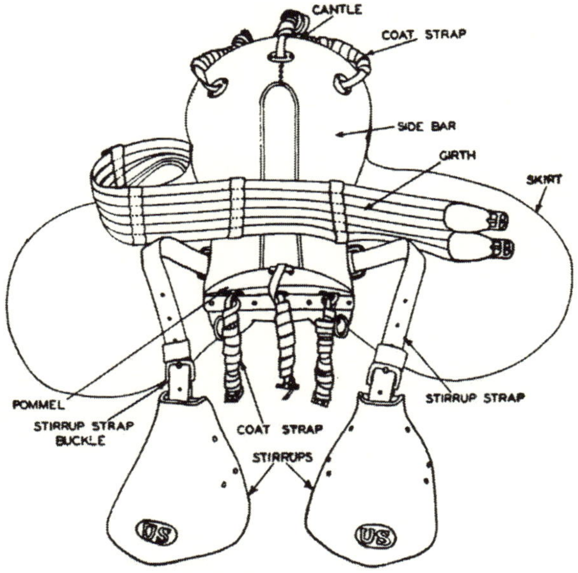

FIGURE 3.—Parts of saddle, McClellan, M1928 (modified), cavalry.

(5) Grasp blanket at the withers with left hand and at the loin with right hand.

(6) Remove blanket in direction of the croup, edges falling together, wet side in.

(7) Place blanket on saddle, folded edge on pommel.

b. Modified McClellan saddle.—When the modified McClellan saddle is used, the instructions given in *a* above govern, substituting the words "unbuckle and let down girth" for "loosen cinch strap and let down cinch."

18. To put on and adjust single or double snaffle bridle.—*a*. The instructor commands: ***BRIDLE***.

(1) Take reins in right hand, crown piece in left hand.

(2) Approach the horse on left side, passing right hand along his neck.

(3) Slip reins over horse's head and let them rest on his neck.

(4) Take crown piece in right hand and snaffle bit in left hand.

(5) Bring crown piece in front of and slightly below its proper position.

(6) Insert thumb into the side of horse's mouth, above the tusk, and press upon lower jaw so as to cause him to open his mouth.

(7) Insert bit by raising crown piece and, with left hand, quietly draw the ears under crown piece.

(8) Secure throatlatch.

b. The bridle should be adjusted so that the snaffle bit will touch lightly the upper corners of the lips and so that the throatlatch will admit four fingerbreadths between it and the throat.

c. At the direction of instructor the halter may be taken off before bridling, the reins first being passed over the horse's neck.

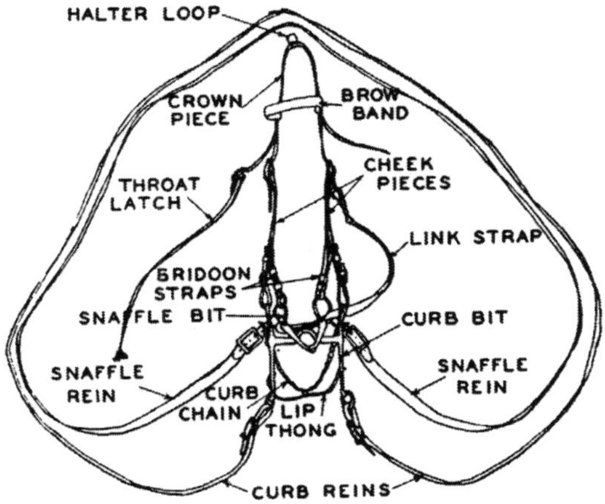

FIGURE 4.—Parts of bridle, M1909, cavalry.

19. To take off single or double snaffle bridle.—*a.* The instructor commands: ***UNBRIDLE.***

(1) Stand on left side of the horse and pass reins over horse's head, placing them on bend of left arm.

(2) Unbuckle throatlatch.

(3) Grasp crown piece with right hand and, assisting with left hand, quietly disengage the ears.

(4) Lower crown piece with right hand and gently remove bit from horse's mouth with left hand.

(5) Place crown piece in palm of left hand and, taking reins in right hand, pass them over crown piece.

(6) The bridle is placed across the saddle on the blanket.

b. If the horse is not equipped with a halter, unbridle and push the bridle back so that the crown piece will rest on the neck behind the poll; or the reins may be left around the neck. The horse may be held thus until a halter is put on, after which the bridle is removed.

20. To put on bit and bridoon.—*a.* The bit and bridoon is put on by the same command and by essentially the same movements as prescribed in paragraph 19 for the single snaffle bridle. Variations in procedure for the two types of bridles are as follows:

(1) Before bridling, the curb chain is unhooked on left side.

(2) After placing the reins over horse's head, the lower left branch of curb bit is held in left hand, forefinger against mouthpiece, snaffle bit above and resting on mouthpiece of curb bit.

(3) After throatlatch has been secured, the curb chain is hooked on left side below snaffle bit.

b. (1) The snaffle should touch lightly upper corners of lips.

(2) The curb should rest just below the snaffle. It should not be placed so low as to strike the teeth.

(3) The curb chain should lie flat in chin groove and be of such length that when the reins are stretched the shanks of the bit form an angle of approximately 45° with lower jaw.

c. When the noseband is used, it should be adjusted to 1 inch below the prominent point of cheekbone.

21. To take off bit and bridoon.—The procedure is the same as prescribed for the single snaffle bridle, except that before removing bits from horse's mouth, the curb chain is unhooked on left side.

22. To stand to horse.—*a.* At the command ***Stand to horse,*** each rider places himself on the left side of the horse, facing to the front and opposite the horse's head. He takes the position of attention, except that the right hand (nails down) grasps the reins about 6 inches from the bit, the forefinger separating the reins. The bight of the reins rests on the horse's neck near the withers.

b. Being at ***stand to horse,*** if the command ***At ease*** is given, each rider stands at ease, holding the reins about 18 inches from the bit.

c. At the command ***Attention,*** each rider resumes the position of ***stand to horse.***

d. To ***stand to horse*** on the right side, the instructor commands: ***RIGHT SIDE, STAND TO HORSE.*** At this command each rider assumes a position on the right side of his horse corresponding to the position of ***stand to horse*** as described above, substituting right for left.

23. To lead out.—The rider being at ***stand to horse,*** to leave the stable or picket line, the instructor commands: ***LEAD OUT.*** Each rider, holding his right hand well up and firm, leads his horse, without looking at him, to the place designated by the instructor.

24. To mount.—The rider being at ***stand to horse*** the instructor commands: 1. ***Prepare to mount,*** 2. ***MOUNT.***

a. At the command ***Prepare to mount,*** the rider faces to the right, drops the right rein, takes a step to the right so as to be opposite the horse's left shoulder, and then half faces to the rear. He takes the reins in the left hand with the little finger between them and the bight falling to the right side; adjusts them so that they give a gentle, even bearing on the horse's mouth, and places the left hand on the horse's crest near the withers, taking a lock of mane, if any, between the thumb and forefinger. He then places the left foot in the stirrup, assisted by the right hand, if necessary, brings the left knee against the saddle, and places the right hand upon the cantle.

b. At the command ***Mount,*** the rider rises by an effort of the right leg aided by the arm. He keeps the left knee bent and firmly pressed against the saddle, the toe depressed, and the upper part of the body inclined slightly forward.

He brings the right foot by the side of the left, changes the right hand to the pommel, passes the right leg, knee bent, over the horse's croup without touching it, and sits down lightly in the saddle. He then places the right foot in the stirrup and takes the reins in one or both hands as instructed.

FIGURE 5.—*Prepare to mount.*

25. To dismount.—Being at a halt the instructor commands: 1. ***Prepare to dismount***, 2. ***DISMOUNT.***

a. At the command ***Prepare to dismount,*** the rider passes the reins into his left hand and places that hand on the horse's crest near the withers. He then places the right hand upon the pommel and removes the right foot from the stirrup.

b. At the command ***Dismount,*** the rider passes the right leg, knee bent, over the croup without touching it. He then changes the right hand to the cantle of the saddle and brings the right foot by the side of the left foot, the left knee being against the saddle and the upper part of the body inclined slightly forward. He next descends lightly on the right foot and takes the position of ***stand to horse.***

26. To hold single reins in either or both hands.—The reins are held, fingers relaxed, knuckles about 30° from the vertical.

a. In both hands.—A rein is taken in each hand in such manner that the rein comes into the hand under the little finger and passes out over the second joint of the forefinger, the fingers lightly closed. The thumb is closed on the rein and the bight falls forward and to the right under the right rein. The hands are well separated and held normally above the withers. The wrist is straight and supple, and the forearm, wrist, hand, and rein form one straight line from point of elbow to horse's mouth, the elbow being slightly in advance of the point of the hip.

b. In one hand.—The reins are held in one hand so that they are separated by the little finger. The two reins pass out of the hand over the second joint of the forefinger, the thumb is closed on the reins, and the bight falls forward and to the right. The free arm swings naturally in rear of the thigh without stiffness.

27. To adjust and handle single reins.—During the first lessons in riding, the instructor causes the rider to hold the reins in both hands. After a few days, however, he requires him to hold the reins in the left hand and, similarly, in the right hand. He practices him in changing the reins from one hand to another, from both hands to one hand and back again, in lengthening and shortening the reins, in dropping and picking up the reins, and explains thoroughly the positions of the hands and the proper length at which to hold the reins.

28. To hold and handle double reins.—*a. In left hand.*—The rider holds the reins in such a manner that the left snaffle rein passes under the little finger, the left curb rein between the little finger and the ring finger, the right curb rein between the ring finger and the second finger, and the right snaffle rein between the second finger and forefinger. He closes his hand so that the reins come out of his hand between the thumb and forefinger, the bight of the reins forward and to the right.

b. In right hand.—The reins are held in the right hand in inverse order to that prescribed for the left hand, the right snaffle rein passing underneath the right little finger.

c. In both hands.—The reins are held in both hands so that the little finger of the hand separates the snaffle and curb reins, the snaffle reins being underneath.

d. Practice in adjusting and handling.—The riders are given practice in changing the double reins from one hand to the

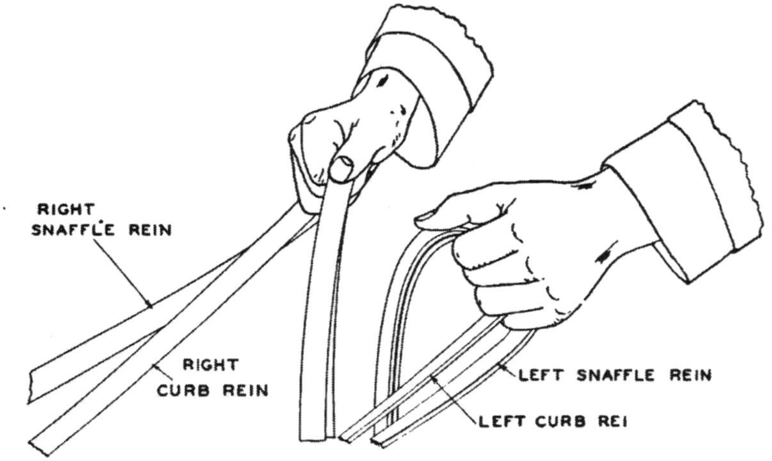

① In both hands.

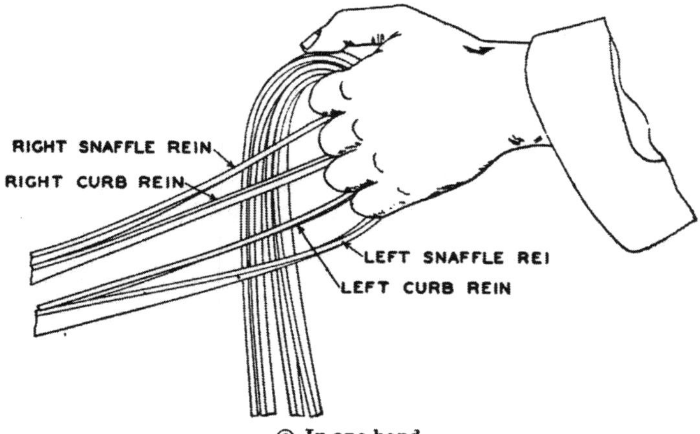

② In one hand.

FIGURE 6.—Proper methods of holding reins.

other, taking them in both hands, taking a single rein in one hand while three are held in the other hand, and in taking the snaffle reins in the right hand and the curb reins in the left hand.

29. Effect of the curb bit.—The effect of the curb bit is much more severe than that of the snaffle because its action occurs almost entirely on the bars of the horse's mouth. The pressure acts downward and tends to lower the horse's head and neck. If the pressure of the curb bit is maintained for a long time, it becomes painful to the horse; also the bars of the horse's mouth become numb, the curb chain stops circulation, and the horse's mouth becomes deadened to any action of the bit. Therefore the curb bit should not be used continuously for long periods. Best results are achieved when the curb is used intermittently and in conjunction with the snaffle bit.

30. Use and effect of bit and bridoon.—The instructor explains and illustrates to the men the following methods of using the bits of the double bridle:

a. The horse should be ridden with a light and equal pressure of the curb and snaffle bits. This is the method used at drill, on the march, and in cross-country riding or jumping.

b. The rider may, on occasion, take the curb reins in the left hand and the snaffle reins in the right. This method is often found useful in the management of a hard-mouthed, stiff-necked horse. If the rider lightly alternates the tension of the curb and snaffle, the sensation thus produced on the bars of the mouth brings about the gradual relaxation of the muscles of the jaw and the poll and, as a result, the horse becomes light and obedient.

c. The rider may also, on occasion, take three reins in the left hand and the right snaffle in the right hand and, by gentle vibrations of the right snaffle rein, cause the horse to relax the muscles of his jaw and poll; or by an upward effect of the right snaffle rein, cause the horse to raise his head when carrying it too low; or by using the right snaffle rein as a bearing rein or leading rein, render assistance to the left hand in turning the horse.

31. To gather the horse.—Having contact with the horse's mouth through the stretched rein and a light feel of the lower legs against his sides, increase the pressure of the lower legs

and slightly increase the pressure of the bit by closing the fingers on the reins until the rider has attracted the horse's attention. Gathering the horse serves as a preparatory signal to attract his attention and to prepare him to respond promptly to the subsequent demand on him through the aids.

32. To move forward.—1. ***Forward,*** 2. ***MARCH.*** Being at the halt, at the command ***Forward,*** gather the horse. At the command ***March,*** incline the weight of the body forward, increase the pressure of the lower legs and relax the fingers of the hand without losing contact with the horse's mouth. If the horse fails to respond, use the legs with repeated applications until the horse moves forward.

33. The slow trot.—1. ***Slow trot,*** 2. ***MARCH.*** Being at the walk, at the command ***Slow trot,*** gather the horse. At the command ***March,*** increase the pressure of the lower legs and at the same time relax the fingers of the hand without losing contact with the horse's mouth. If the horse fails to respond, use the legs with repeated applications until the horse takes the slow trot. The rider then maintains the 6 m. p. h. rate by using his legs to preserve the forward movement and, by means of the direct reins, to control and regulate this movement. The rider sits in the saddle, the upper body erect, all muscles and joints relaxed, the hands low and light.

34. The trot (posting).—1. ***Trot,*** 2. ***MARCH.*** At the command ***March,*** the rider increases the gait by the same methods as for the slow trot, except that the rate is increased to 9 m. p. h. The rider's body moves upward and forward, then downward and backward in cadence with the beats of the gait. Sufficient forward inclination of the upper body is taken for the rider to be in balance. The upper body maintains its posture unchanged without sinking rearward at the loins, as the rider comes into the saddle at each alternate beat of the trot. The chin is raised so that the plane of the face remains vertical. The rider sinks into the saddle very lightly on the upper thighs and crotch, the points of the buttocks barely touching the saddle at each beat.

35. The gallop.—1. ***Gallop,*** 2. ***MARCH.*** Being at the slow trot or trot, at the command ***Gallop*** (right), apply the aids for the gallop. The rider increases the pressure of both legs, placing the left leg in rear of the girth and the right leg on the girth.

At the same time he increases the tension of the right rein to the left and to the rear, and the tension of the left rein lightly to the rear. This frees the right shoulder and restrains the left. At the command ***March,*** increase the action of both legs and lessen the pressure on the reins sufficiently to allow the horse to move forward at the gallop. At the gallop the rider's upper body is inclined forward slightly more than at the walk or slow trot.

36. To halt.—1. ***Squad,*** 2. ***HALT.*** At the command ***Squad,*** gather the horse. At the command ***Halt,*** use the direct reins and the weight in combination until the horse halts. In using the weight, the rider should keep the crotch deep in the saddle, the legs in their proper place and snugly against the horse.

37. To rein back.—Being at the halt the instructor commands: 1. ***Backward,*** 2. ***MARCH.***

a. At the command ***Backward,*** the rider gathers his horse.

b. At the command ***March,*** the rider closes his fingers on the adjusted reins, at the same time shifting the weight slightly to the rear. The action of the reins should not be steady or continuous, but is accomplished with quick, repeated applications until the horse takes one or two steps to the rear. After the horse commences his first step to the rear the fingers relax and close again as the next step is demanded. The legs act in keeping the horse straight and in counteracting any tendency to back too fast.

c. To halt, the action of the direct reins should cease, the weight carried forward, and the legs used sufficiently to stop the backward movement.

38. To change hands.—This movement is used only in a riding hall or on a riding track. At the command ***Change hands,*** each rider, after having passed the corner of the riding hall or track and marched a horse length on the long side, directs himself toward the diagonally opposite corner so as to take the track to the opposite hand at about two horse lengths from the corner. Those going in an opposite direction are passed by keeping to the right. If in column the leader executes the above movement, those in rear follow in trace.

39. By the right (left) flank.—1. ***By the right flank,*** 2. ***MARCH.***

a. Being at the halt, at the command ***By the right flank,*** gather the horse. At the command ***March,*** move the horse forward, apply the leading rein (or the bearing rein) in such a manner as to turn the horse to the right, at the same time closing the right leg on the girth and the left leg slightly in rear of the girth. The turn is made on an arc of 90°. The radius of the arc at the walk should be 2 yards; at the trot, 4 yards; at the gallop, 6 yards. During the movement the body of the horse should follow the arc on which he is turning. If the horse carries his haunches to the outside of the curve his balance is adversely affected, and the rider should oppose such action by using the left leg more strongly and slightly farther back than the right leg. In the first lesson in turning by the flank the rider uses the leading rein. Later he is taught to use the leading rein, assisted by the bearing rein; then the bearing rein alone, first with the reins in both hands then in one hand.

b. ***By the left flank*** is executed by similar commands and means.

40. Right (or left) oblique.—This movement is executed in the same manner as ***by the right or left flank,*** except that each rider turns his horse through an arc of only 45°. The command is: 1. ***Right (or left) oblique,*** 2. ***MARCH.*** To resume the original direction, 1. ***Forward,*** 2. ***MARCH.***

41. The about.—At the command 1. ***Right (or left) about,*** 2. ***MARCH,*** each rider turns his horse until he faces in the opposite direction and then moves to his new front.

42. Changes of gait.—*a.* Being at the walk, to pass to the trot, or being at the trot to pass to the gallop, the rider uses the same means as for passing from the halt to the walk. Likewise, being at the gallop to pass to the trot, or being at the trot to pass to the walk, the rider uses the same means as for passing from the walk to the halt.

b. Changes of gait are habitually made progressively, the walk being taken from the halt, the trot from the walk, the gallop from the trot, and vice versa. For instruction purposes and when necessary in service an increase or decrease of gait may be made more abruptly.

c. A change of gait of one degree means a change from the halt to the walk, from the walk to the trot, from the trot to the gallop, or vice versa.

43. To charge individually.—*a.* In preparing the cavalry troopers to participate in the charge, the instructor should require exercises in gradually extending the gallop and in galloping individually at full speed over a distance of 60 to 75 yards. In charging, the rider should bend the body forward and push his feet home in the stirrups. At the beginning of the charge, the rider shortens his grasp on the reins so as to enable him better to control his horse. He should close his legs firmly on his horse, using the spur if necessary, and hold him in the direction indicated. Having charged the distance prescribed by the instructor, the rider should reduce the speed to the regulation gallop.

b. In order not to fatigue the horses by repeated charges, the individual charge should be made a part of several instruction hours. It should always be executed on a piece of ground especially adapted to this work; that is, open, level ground where the footing is neither too hard nor too soft.

44. To rest.—*a.* Being at ***stand to horse,*** the command ***Rest*** is executed as when dismounted without arms, except that the men hold the reins and keep their horses in place.

b. Being mounted, at the halt, at the command ***Rest,*** or being in march, at the command ***Route order,*** the men are permitted to turn their heads, to talk, to smoke, and to make slight changes of position, but not to lounge in their saddles or to lose their relative places.

c. Being at ***stand to horse,*** the command ***At ease*** is executed as when dismounted without arms.

d. Being mounted, at the command ***At ease,*** the men are permitted to ride as in route order, except that they preserve silence.

45. To resume attention.—The instructor commands: 1. ***Squad,*** 2. ***ATTENTION.*** At the command ***Attention,*** each rider, if dismounted, takes the position of ***stand to horse,*** at attention; if mounted he takes the position of ***attention*** mounted.

46. Suppling exercises.—*a.* Suppling exercises have as their objective the general suppling and strengthening of the rider's muscles and joints, independence and control in the employment and reactions of his various parts, familiarity with the horse, and increased confidence. The mounted suppling exercises, like all physical culture exercises, depend for their good results upon the regularity and thoroughness with which they are practiced for a short time daily. To produce the best results, these exercises must be performed correctly. Before taking up a new exercise, the instructor gathers the class about him and explains and illustrates the exercise. He then causes them, at the halt, to practice the exercise just explained.

b. The gaits employed during the suppling exercises should be the walk, the slow trot, and the gallop. The gallop should be employed very early in the instruction and used chiefly thereafter, because it is the most favorable gait for suppling the rider's loins. The slow trot is trying to inexperienced riders and, if prolonged, may cause such abrasions of the skin and soreness of the muscles as to retard progress. However, there should be much work at the slow trot, but for short periods only.

c. The instructor causes the riders to take the track without regard to distance; if they march to the right, they hold the reins in the left hand; if to the left, they hold the reins in the right hand. The instructor may cause the riders to knot the reins and release them entirely.

d. The exercises are not performed in unison. While practicing an exercise the riders should maintain the correct position and steadiness of the bridle hand and the suppleness of the wrist, elbow, and shoulders. He should endeavor to coordinate and to separate his movements. The exercise is continued by the riders until the command ***As you were*** is given, whereupon each rider ceases the exercise and resumes the initial position.

e. The instructor should give his commands, observations, and criticisms in such a manner as to be easily heard and understood. When prolonged or special explanations are necessary, the rider in question should be required to fall out of the column and report to the instructor. If several riders are involved, the instructor should assemble the whole column in order that all may profit from his instructions.

f. The following exercises are employed:

(1) *To supple the neck.*—The object of neck exercises is to combat rigidity of position due to the contraction of muscles in the region of the neck.

(*a*) The instructor commands: ***ROTATE THE NECK.*** Each rider slowly turns the head and eyes as far as possible, first to the right, then without pause to the left.

(*b*) The instructor commands: ***FLEX THE NECK.*** Each rider slowly carries the head as far as possible upward and backward, then, without pause, as far as possible forward and downward.

(2) *To supple the shoulders.*—The object of shoulder exercises is to combat rigidity of position due to contraction of muscles in the region of the shoulders and to cultivate coordination of muscular effort. The instructor commands: ***ROTATE RIGHT (LEFT) ARM VERTICALLY.*** Each rider extends the arm straight upward, wrist and hand relaxed; he then carries the arm to the rear, downward, forward, and upward in such a manner as to describe, with uniform movement, a vertical circle.

(3) *To supple the loins.*—The object of these exercises is to develop suppleness in the region of the loins.

(*a*) The instructor commands: ***SWING RIGHT (LEFT) ARM HORIZONTALLY FROM FRONT TO REAR.*** Each rider extends the arm straight to the front, fingers extended, palm upward. He then, by rotating the body in the region of the loins, swings the shoulders and arm to the right, the arm moving in a horizontal plane until the hand is over the horse's croup, the eyes following the hand.

(*b*) The instructor commands: ***STROKE THE HORSE ON RIGHT (LEFT) SHOULDER WITH RIGHT (LEFT) HAND.*** Each rider, by flexing the body at the waist, leans slightly forward and downward and gently strokes the horse on the shoulder.

(*c*) The instructor commands: ***STROKE THE HORSE ON LEFT (RIGHT) SHOULDER WITH RIGHT (LEFT) HAND.*** Each rider, by rotating the body in the region of the loins, carries the hand over the reins and downward to a position opposite the designated shoulder of the horse. He then gently strokes that shoulder.

(*d*) The instructor commands: ***STROKE THE HORSE ON LEFT (RIGHT) SHOULDER AND RIGHT (LEFT) HAUNCHES WITH RIGHT (LEFT) HAND.*** Each rider, in the same manner as in preceding exercises, strokes the horse on the designated shoulder with the designated hand and then, without pause, while rotating the body in the region of the loins, turns the shoulders and arm to the right (left) rear, the eye following the hand, and strokes the horse on the designated haunch. He continues the exercise, alternately stroking the horse on the shoulder and haunches.

(*e*) The instructor commands: ***SWING LOW TO THE FRONT WITH RIGHT (LEFT) HAND.*** Each rider extends the arm vertically upward, palm to the front, leans to the right (left) and downward by flexing the body at the waist, and makes a low sweeping stroke from the rear to the front; he then resumes the initial position of the body, the arm extended vertically upward. With a slight pause, he thus continues the exercise.

NOTE.—In the preceding exercises the lower body should be kept in its proper position in the saddle.

(*f*) The instructor commands: ***RAISE THE THIGHS.*** Each rider, inclining the upper part of the body slightly backward, raises the thighs only so much as necessary to detach them and the lower legs from contact with the saddle. The rider thus balances on the buttocks, which should be pushed well underneath the body. This exercise requires the rider to flex the body at the loins in order to follow the motion of the horse and to minimize the effect of the shock. Riders should not be required to maintain this position continuously; they take and retake it, according to their ability to do so without falling.

(4) *To supple the hip joints.*—The object of this exercise is to supple the muscles in the region of the hips and flatten the inner thigh surfaces of the rider's legs. The instructor commands: ***ROTATE RIGHT (LEFT) THIGH.*** Each rider removes the thigh from against the saddle, straightens the leg, and carries it back until vertical. He then rotates the thigh at the hip joint by turning the knee inward and, pressing the thigh hard against the saddle, draws it forward to its normal position. The lower leg is replaced in its normal position. The foot should hang naturally without constraint. This exercise should habitually be given at the halt; if executed while the horse is in motion, serious injury may occur in the region of the hip joint. The riders being at the halt, the instructor directs them to rotate the right thigh and carefully fix it in

position; then, similarly, the left thigh. He then causes the riders to take the walk, the slow trot, or the gallop, during which they should try to confirm the new position. After a few moments, however, the riders become shaken out of the exact position assumed. The instructor then brings them to the halt, directs them to rotate again one thigh, then the other, and to fix their position, after which he causes them again to move forward.

(5) *To supple the knee.*—The object of this exercise is to render the knee joints supple, combat the involuntary contractions to which the knee joints are liable, and cultivate the coordination of muscular effort. The instructor commands: ***FLEX RIGHT (LEFT) KNEE.*** Each rider, removing the lower leg from against the horse, carries it backward and upward until horizontal, then without pause returns it to the normal position.

(6) *To supple the ankles.*—The object of these exercises is to assist the rider in overcoming involuntary contraction and stiffness in the ankle joints, in lowering the heels properly when riding with stirrups, in retaining the stirrups by the easy flexibility of the ankle joints, and in overcoming the tendency to stand stiffly in the stirrups.

(*a*) The instructor commands: ***ROTATE THE RIGHT (LEFT) FOOT.*** Each rider traces with the toe, by a slow and uniform movement, a circle in a vertical plane perpendicular to the horse. In tracing the circle the rider's toe first moves toward the horse. The exercise is thus continued.

(*b*) The instructor commands: ***FLEX THE RIGHT (LEFT) ANKLE.*** Each rider slowly flexes the ankle by raising the toes upward as far as possible then, without pause, extending them downward as far as possible. The exercise is thus continued.

47. To ride without stirrups.—Riding without stirrups is a very useful exercise to develop a strong seat. It should never be employed until the recruit has gained confidence in his ability to direct and control his horse, and has learned to post and gallop with the stirrups. In riding without stirrups, with the ankles relaxed, the toes normally will be lower than the heels.

48. To ride across country.—*a*. During the first period of instruction in cross-country riding, the unevenness of the ground selected should not be great. All fences, ditches, and artificial obstacles should be avoided. The gait during the first few lessons is the walk, with frequent halts to vary the monotony. Later other gaits are added. The men should sit at their ease without lounging in the saddle.

b. A ride out-of-doors should be further diversified by changes in formation. Starting out habitually in column of twos the instructor should soon scatter the riders out in a flock so as to cause each man to ride as an individual. The speed of the regulation gaits should be maintained. Abrupt changes of pace or sudden bursts of speed may cause confusion.

49. To jump obstacles.—The jumping of obstacles is an excellent exercise for developing a good seat and leads to boldness, confidence, and aggressiveness in the rider and horse. Jumping may be analyzed into three parts: first, conducting the horse straight upon the center of the obstacle; second, inspiring him with sufficient impulsion to insure his jumping; and third, going over the obstacle with the horse in such a manner as not to interfere with his balance or freedom of movement at the take-off, in the air, or on landing. The correct muscular reactions of the rider in the third operation are called the "gymnastics of the jump." The first two parts may be considered together and constitute the approach, or conducting the horse to the obstacle. The beginner must first attain some degree of proficiency in the gymnastics of the jump before adding the problem of conducting the horse to the obstacle.

a. Gymnastics of the jump.—The gymnastics of the jump develop in the rider balance or the ability to go "with" his horse, coordination and independence in the use of the parts of the body, suppleness, muscular control, security of seat, and confidence. Initially, well-trained, willing, and quiet horses are used over the simplest of obstacles. It is advisable to start at first with bars placed on the ground and gradually raise them by degrees, about 3 inches at a time, until the riders can take without difficulty an obstacle 2½ feet in height. The following are the essential points in the correct jumping seat:

(1) The rider approaches the jump seated, but with body inclined forward. He has contact with the saddle along the full length of his thighs from knees to points of buttocks. His

feet are well home in the stirrups. His weight is borne principally by the thighs, diminishing rearward until at the points of the buttocks it is practically nil. His body during the approach, while still seated in the saddle, has sufficient forward inclination to insure its center of gravity being in line with the direction of the horse's thrust and slightly in advance of the center of its base of support. If the forward inclination is not sufficient, or if the back is humped (not straight), the line of thrust passes beneath the center of gravity of the

FIGURE 7.—The approach.

rider's body causing him to rock to the rear or break badly at the loins. As a result the rider is "behind his horse" or, as is frequently said, "his horse jumps from under him." It must be remembered that whenever the body is inclined forward, the inclination is made from the hips and not at the loins.

(2) During the take-off the hands are absolutely passive; that is, they follow the mouth with a constant even tension on the reins. Any unnecessary movement of the hands or shifting of the rider's weight distracts the horse and upsets his balance.

(3) If the rider's position is correct at the take-off, he is thrown slightly forward and upward without voluntary effort.

FIGURE 8.—The take-off.

FIGURE 9.—Over the jump.

The knee angles open and the hip angles close. The angles of the ankles likewise close due to the increased weight in the stirrup.

(4) Over the jump the buttocks are kept out of the saddle by the rider's balancing on his lower thighs, knees, and stirrups. The back must not be too relaxed, as muscular control of body and balance would then be lost, causing the rider to fall back into the saddle or forward on the neck of his horse.

(5) During the descent and in landing, the forward inclination of the body is gradually decreased to avoid losing balance, falling forward on the horse's neck, or an abrupt return of the buttocks to the saddle. The hip angles open, the knee angles close, and the rider is preparing to resume his seat.

(6) On landing, the shock is progressively received on hip joints, thighs, and knees, and through the ankles onto the stirrups. As the horse's forefeet are grounded and his hindquarters come to the relief of his forehand, the rider's body sways forward from the hip joints to save the horse's back from unnecessary strain. The body is prevented from falling forward by the correct amount of inclination, assisted by muscular efforts of the back, lower thighs, knees and legs, and by keeping the posture of the loin and back unchanged. Head and chin are kept high. For a few strides the rider continues to support himself on the lower thighs, knees, and stirrups; then he relaxes the knees and settles gently into the saddle.

(7) The rider's crotch is deep in the saddle, fleshy part of the buttocks to the rear and not under the rider.

(8) Thighs, knees, and heels are forced down to the limit allowed by the length of the stirrup straps.

(9) The legs are always against the horse and not in cadence with the stride of the horse.

(10) Head, chin, and eyes are up. Holding the chin up assists greatly in maintaining correct spinal posture and muscular control of the body.

b. To conduct the horse to the obstacle.—(1) When the rider has made sufficient progress in the gymnastics of the jump, the difficulties of the approach may be taken up. Either more difficult obstacles may be used or two or more obstacles placed one after the other (in and out). Another method is to remove the wings. Care must be taken to make the work progressive, and to use obstacles of no greater height or breadth than the horses can negotiate.

(2) As the horse starts his approach, often somewhat abruptly, the rider inclines his body forward sufficiently to keep his balance. Thrusting the body forward or sinking backward at the loin at the moment of take-off must be avoided.

(3) The approach should be made on a line perpendicular to the obstacle and straight toward the center. The speed is regulated according to the size of the jump and upon the natural "manner of going" of the horse. Speed should be

FIGURE 10.—Landing.

gained gradually as the jump is approached. The rider should begin the approach not too far distant from the obstacle, 15 or 20 yards, with the horse "on the hand," gathering impulsion and only the necessary amount of speed as he approaches, and with a reserve of impulsion to be used at the last moment should the horse hesitate. Rushing or excessive speed must be avoided. The reins should not be too long, so that any attempt of the horse to run out may be forestalled; hands low and well separated; and the horse evenly balanced between hands and legs. As the approach begins, the fingers slightly relax on the reins, and the legs slightly increase their pressure, requiring and facilitating a gradual increase of speed and impulsion. The rider must determine and use the necessary aids to conduct his horse straight and to insure his jumping. The legs should not be "thrown away" on a horse which rushes a jump. Such a horse is just as likely to refuse, and more likely to run out, than the quiet-going horse. The rider should feel or sense the horse during the approach, and use his legs sufficiently to insure jumping. He should approach the obstacle determined to jump.

(4) During early instruction in jumping, the curb reins should be knotted and dropped on the horse's neck or fastened to the throatlatch of the bridle.

(5) When riding across country, only obstacles up to the ability of horse and rider are taken.

50. To swim horses.—*a.* The first and most important step is to teach the men to swim well. A poor swimmer may render useless the efforts of a horse that can swim well. On the other hand, nearly all horses will swim if accompanied by a man who swims easily.

b. Before sending men and horses into the water for practice, at least two men who are strong swimmers are posted as life guards on each side of the stream or pool. If available, a boat or raft is located conveniently for picking up men who get into difficulty.

c. In teaching men to swim horses, the halter with shank and snaffle bridle are used. The halter shank is tied around the horse's neck and the snaffle reins are knotted to prevent becoming entangled in the horse's legs.

(1) The rider directs his horse into the water by means of the reins. When the horse commences to swim, the rider grasps the halter shank which has been tied around the horse's neck.

(2) The proper position of the rider to interfere least with the horse's balance is approximately the position of the trooper mounted, but with the upper body inclined slightly forward to meet the slight up-thrust of the front part of the horse. It is necessary for the rider to grip considerably with his legs.

(3) The rider's weight carried too far forward tends to lower the horse's head in the water and push it under. The rider's weight carried too far to the rear tends to sink the

haunches farther under water and cause the fore hand to come up, thereby impeding forward progress.

(4) The rider guides the horse by a slight tension on the reins (used as a leading rein). He should avoid splashing water on the horse's head and face as it tends to excite the horse.

d. The following alternate positions of the rider in swimming horses are sometimes used:

(1) After the horse has started swimming in the proper direction, the rider slides off the horse's back on the downstream side and floats or swims alongside the horse, keeping one arm over the horse's withers with a grasp on the halter shank and using the other arm for swimming or for grasping a rein to lead the horse in the proper direction.

(2) After the horse has started swimming in the proper direction, the rider slides over the horse's croup, grasps the tail, and is towed by the horse. A knot tied in the tail will assist the rider in grasping it. To direct the horse, the rider splashes water on the side from which he wants to turn, being careful to avoid splashing water on the horse's head or face.

e. When teaching horses to swim, care must be taken not to frighten them. All horses can swim, and it is only on account of fear that they sometimes become difficult to manage. Several bold swimming horses should be used to lead the others into the water. A few of these sent across to stand on the far bank will encourage the others to cross and land. Horses that at first refuse must either be led beside a bold swimmer or be towed across from a boat or raft. In the latter case a tow rope must be used, as oars will frighten a horse.

f. When all of the horses of an organization have learned to swim a stream, they may then be sent across in a herd, led by a few steady horses with their riders. When several groups cross in succession, a few of each group should be left on the far bank to attract the horses of the succeeding group.

g. Loaded pack animals may turn over and drown, especially if there is much current. When time and material are available, small boats or rafts should be used to carry over the clothing, arms, equipment, saddlery, and pack loads, the men swimming with their horses. In cold weather the horses should be herded across and the troopers with their equipment and saddlery taken over in boats or rafts.

h. If it is necessary for the horses to swim with saddles or carrying arms and equipment, accompanied by the troopers, the following precautions must be taken:

(1) Curb reins must be wrapped around each other and secured by the throatlatch, snaffle reins knotted, stirrups crossed, girths adjusted snugly, and no strap or other article left in such position that man or horse can become entangled in it. Ammunition belts and pistols must be attached to pommels of the saddles, spurs removed, canteens emptied and tightly corked.

(2) Officers and noncommissioned officers supervise the preparations for the crossing to insure that all safety precautions have been observed by each trooper before he enters the water. Pack loads must be carried over in boats, rafts, or some other improvised floats. Nonswimmers who cannot be carried otherwise must grasp the tails of steady horses. Knots should be tied in the tails to prevent hands losing their hold.

(3) Before swimming a river, reconnaissance must be made to determine the most favorable crossing. When necessary the banks will be prepared to facilitate going into and out of the water. At night the entrance and exit should be lighted by lanterns. When the current is swift, an additional landing should be prepared about 200 yards downstream for the benefit of any horses that might be swept past the regular landing.

CHAPTER 2

REMOUNT TRAINING

SECTION I

GENERAL

51. Object.—*a.* The object of remount training is to develop, in a minimum of time, the condition and training of the remounts to a degree necessary to fit them for field service in the ranks of the troop.

b. A good troop horse is one that is sound and of good conformation, in good condition for hard work, tranquil, easy to ride and manage, a moderately good jumper, unafraid of water and a good swimmer, and accustomed to the use of arms and the sound of firing.

c. In the training of remounts, great attention should be paid, first, to their conditioning; second, to their tranquility; third, to their training. *Any system of training that neglects the conditioning or destroys the tranquillity of the remounts is defective.*

52. Physical fitness.—At the beginning of their training, remounts are not in good physical condition. Therefore the daily work imposed upon them should tend to be long in point of time but very mild in character. There should be frequent rests during which the riders look their horses over carefully, examine their feet, and readjust equipment. Long periods at the walk, very short periods at the trot, and shorter and less frequent periods at the gallop expand the lungs, strengthen the heart, and build up the muscles of the new horses. The instruction should be progressive, patient, and moderate. The training in exact obedience to the aids and in jumping obstacles should be delayed until gentling and conditioning are well under way.

53. Organization into suitable groups.—Not less than 8 nor more than 32 remounts should be trained by one instructor at one time. The remounts are organized into squads and are thus accustomed to simple military formations and evolutions while being trained.

54. Selection of riders.—*a.* For the training of remounts those men should be selected who have a liking and an aptitude for the management of animals and who, in addition, have received a thorough course of instruction in riding.

b. In the assignment of riders to horses some adjustment will be necessary during the first few days in order to place certain types of horses in the charge of men who appear best able to manage them. As soon as practicable, however, all assignments should be made permanent.

c. In the early lessons, especially when out of doors, the officer in charge should designate as group leaders those men who are riding the oldest and gentlest remounts.

55. Care of remounts.—*a.* The officer in charge should pay scrupulous attention to the cleanliness of stables, stalls, and mangers, and to the general sanitation of the immediate surroundings. Remounts are very susceptible to colds, fevers, and other ailments which will interfere with or interrupt the training. He should, by daily inspections, keep himself thoroughly informed as to the kind and supply of forage, its preparation, the amount fed, and the frequency of feeding. (See ch. 3.)

b. During work the officer in charge must be watchful of the rate and duration of the various gaits to prevent overtaxing the strength of the remounts. No great attempt should be made to conform to the regulation rate at the walk, the trot, or the gallop until all the remounts have attained a fair degree of condition. Thereafter the regulation rate of the various gaits should be maintained.

c. As rapidly as possible the remounts should be shod, beginning with the quietest ones first. The officer in charge should supervise the shoeing to make sure it is properly done in a quiet, gentle manner.

56. Fitting of bits and saddles.—*a.* Ill-fitting equipment may seriously injure the remount's temper. Accordingly the proper fitting of bits and saddles should receive systematic attention. (See sec. II, ch. 1.)

b. In fitting the snaffle bit in the remount's mouth it is very important to prevent the horse's tongue from getting over the bit. Therefore the cheek pieces of the bridle should be carefully adjusted in length so that the snaffle bit will touch lightly the upper corners of the lips.

57. Duration of training course for remounts.—*a.* The training given remounts consists of conditioning or physical upbuilding, preservation of tranquility or good disposition of the horse, and the training proper, or "schooling." The time required under each of the above headings is subject to wide variations due to such factors as season of the year; the age, quality, and condition of the remounts; the facilities available, such as riding halls, terrain for cross-country riding, jumping pens, etc.; ability of the riders; and amount of time available each day for the work. Except in an emergency, 3 or 4 months should be employed in the training of the average remount before assigning it to troop duty.

b. The conditioning of the remount requires more time than does the other training. No matter how great the rider's skill, the physical development of the horse is mainly subject to the laws of nature. Increased work cannot hasten the natural evolution and substitute itself for the effect of time.

c. Due to sudden war emergency, it may be necessary to prepare remounts for field service in a shorter period of time than 3 or 4 months. The training outlined in this chapter is based on the existence of such an emergency, and it must be clearly understood that the condition of the remounts at the conclusion of such brief training will not attain the standard desirable for field service. Horses so trained must necessarily receive a large share of their conditioning after assignment to troop units. In particular the weight carrying muscles of such horses must be carefully strengthened and hardened. This may be accomplished by requiring a large amount of leading in the early stages of campaign, and a very gradual increase from day to day in the time the rider's weight is allowed to be on the horse's back. Unless the strength of the horse's back is carefully built up, a large percentage of the new horses will be quickly incapacitated for further service.

d. In conducting the training of remounts under the conditions specified in *c* above, the officer will find it advisable to segregate the younger animals and those of poorer quality and substance and have their training proceed with greater deliberation.

58. Division of training into periods.—The course of training herein prescribed covers 8 weeks in time, divided into four periods of 2 weeks each. These periods may be shortened or lengthened, but the subject matter of which each period consists should always be covered, either hastily if necessity demands, or more deliberately and thoroughly if time and conditions permit.

59. Weekly inspection of horses.—The officer in charge of the training of remounts should inspect each horse in a formal manner at least once each week. He should then question the rider and discuss with him the condition of the horse. The two together should minutely inspect the teeth, eyes, nostrils, feet, legs, coat, and should note the state of health of the horse. They should closely observe for any indication of approaching diseases or lameness. Where possible a veterinarian should also be present at this inspection.

60. Periodic inspection by commanding officer.—At the end of each period the commanding officer should inspect the remounts and should test the results accomplished in their conditioning and training. In his tests he should confine himself strictly to the subject matter of the approved program of instruction.

Section II

FIRST PERIOD

(First 2 weeks)

61. Program of daily training.—*a.* When local conditions permit, the gentling and conditioning of remounts should be started while they are still in quarantine. Healthy remounts should receive such amount of exercise as the veterinarian recommends and of such nature as the facilities available and the provisions of the quarantine will permit. As a general rule, exercise should be confined to herding in corrals or on an exercise track in the quarantine area. Great care should be taken to see that they do not become heated or fatigued. During quarantine, remounts must recover from the fatigue of shipping and should gain weight and strength in preparation for the active conditioning and training which will follow. Each healthy animal should be groomed once each day, preferably at a feeding time.

b. After release from quarantine the remounts should receive two training lessons per day, one in the morning and one in the afternoon, each lesson of about 1½ hours' duration.

c. After each lesson the remounts are thoroughly groomed. During the grooming the feet are picked up and the bottoms of the feet tapped with the currycomb or a stone in order to prepare the remounts for their first shoeing experience.

d. The first 30 minutes of each lesson should generally be used for exercising the remounts by herding them around on the exercise track or in a corral, or by leading them beside old horses. The next 45 minutes of each lesson should generally be conducted in a riding hall or inclosure. Any time remaining should be used in riding the remounts across country or on the roads, leading, gentling, picking up feet, and grazing when possible.

e. The remounts are equipped with the single snaffle bridle with single reins and stripped saddles. The troopers ride without spurs and may carry a riding whip or switch when necessary.

f. The exercises outlined in the following paragraphs, in the order in which listed, constitute the subject matter from which the officer in charge should make up his program of daily training.

62. Conditioning.—*a.* Generally not more than 45 minutes should be given to riding hall work at one time. Too much time spent in work in the riding hall dulls the sensibilities of the remounts and they lose that alertness of manner and vigor in action which conditioning should cultivate. Exercise out of doors should keep pace with the exercises and movements required in the riding hall.

b. Condition of remounts is based upon a great amount of walking in proportion to the time spent at the trot or gallop.

During the first 3 days' work the walk only should be used while riding remounts. Thereafter short periods at the trot and gallop may be interposed with long periods at the walk.

c. The rate of the walk must be regulated on the speed of the average horse so as to avoid causing the slower horses to fret and become upset. In returning to the stables, a different route from that used in going away from the stables is followed. The remounts should be brought back to stables dry and cool.

63. To exercise remounts.—*a.* The remounts are exercised by herding them around an oval-shaped track which has been fenced on both sides. To construct this track, any available corral or riding pen of suitable size may be used for the outer fence. Approximately 10 feet from the outer fence an inner fence about 4 feet high is built. Gates should be provided in both fences. A mounted soldier as leader and one or more as drivers are employed to keep the remounts on the move while being exercised. The horses should be driven quietly and, particularly during the first exercises, slowly, so as to prevent their becoming excited and nervous. Shouting and the cracking of whips are carefully avoided.

b. If time and other facilities permit, the remounts may be exercised by leading them beside old horses or by longeing. When plenty of time is available, leading is a more desirable way of exercising them than by herding on the track because it develops quietness and regulation gaits. To quiet a remount before saddling or to discipline and exercise the more unruly ones, longeing may be resorted to when time and equipment are available.

c. As soon as the remounts can be saddled they are required to carry the saddle during all exercise work.

d. Beginning as soon as the remounts move around the exercise track smoothly and without confusion, two or three rails or small logs are placed on the ground across the track at convenient intervals apart. After they have passed over these rails a dozen times, first at the walk and later at the trot, the rails are removed. Thereafter on each succeeding day the rails are raised a few inches at a time until, by the end of the second week, the remounts have been practiced in jumping obstacles about 2 feet high.

64. To put on bridle and saddle.—For the first lessons in saddling and thereafter until the remount submits quietly to saddling, the following procedure is observed:

a. Starting with the first day, all remounts that can be handled are saddled.

b. Before being saddled, the remount is thoroughly exercised by herding on the exercise track, by leading or by longeing.

c. In bridling and saddling, the riders work in pairs, and the remount to be saddled is led away from the other horses. Here the snaffle bridle is put on, care being taken to have the length of the cheek pieces such that the snaffle bit just wrinkles the corners of the horse's mouth. For instructions in putting on the single-snaffle bridle, see paragraph 18.

d. In saddling, the assistant stands squarely in front of the remount, grasping the bridle reins about 6 inches from the bit with the right hand. With his free hand the assistant strokes the remount's head and neck and uses assuring tones of voice to keep him calm. The rider now approaches the remount from the left front with the saddle blanket folded and resting on the right arm; he runs his left hand along the horse's crest and withers, strokes him on the shoulder and back, and places the blanket quietly in place. The rider next places the saddle on the horse's back in the same manner as prescribed for the blanket. If necessary the assistant holds the blanket in place until the saddle is placed on it. The girth is adjusted comfortably; it is not drawn tight. The stirrups are removed from the saddle before attempting to place it on the remount.

e. The remount is then led about for a few minutes and the saddle pressed and handled. As soon as the horse is quiet, the bridle is removed and the remount turned loose on the exercise track. The pair of riders now saddle another remount, following the same procedure.

65. To stand to horse, lead out, and line up (in the riding hall).—*a.* The remounts having become accustomed to the weight of the saddle, the pressure of the girth, and the movement of the swinging stirrups, they are ready for further training in being handled by the rider on foot. For this purpose each rider takes the position of stand to horse and leads his remount up and down at the walk. The riders are next required, while leading their remounts, to line up (in the riding hall), each rider at stand to horse with his horse properly disposed. The first lesson should cease here and the men directed to fall out and lead their remounts to the stable.

b. In order to avoid accidents due to kicking, each rider must keep his horse's head up while leading him.

66. To mount and dismount.—*a.* On the second day the remounts may be ready to be mounted. They are first exercised on the herding track, then saddled, and again returned to the herding track for more exercise. After a thorough workout on the track, three or four of the quietest remounts are brought off the track to the inside of the pen and bridled with single-snaffle bridles. The cavesson and longe instead of the bridle may be used to advantage in the first mounting lessons.

b. The officer in charge designates four or five of the most experienced men to assist the riders in mounting their remounts for the first time or personally assists each rider in this first mounting lesson. The assistant stands in front of the remount, grasping the bridle reins about 6 inches from the bit with the right hand. By his firm, gentle manner and soothing tones of voice the assistant attracts the attention of the remount and keeps him quiet. The rider, meanwhile, approaches the horse from the left front, runs his left hand along the horse's crest and withers, and with the right hand strokes the horse's shoulders, back, and haunches. He places his left hand on the horse's neck just in front of the withers, extends his right arm across the middle of the saddle, and presses down upon the saddle. He then takes the reins in the left hand and places this hand on the horse's neck just in front of the withers, puts the left foot in the stirrup without touching the horse's side with his toe, places his right hand upon the cantle of the saddle, and rises slightly on the stirrup by an effort of the right leg, the left knee bent and pressed against the saddle. If the remount does not move, the rider mounts without hurry, but without hesitation, and sits down lightly in the saddle.

c. If the horse moves when the rider starts to mount, the assistant takes hold of the cheek pieces of the bridle and by stroking or patting the horse induces him to stand still. The assistant must not pull on the reins; to do so may cause the horse to rear and fall over backward. The rider should not mount the horse until he stands still again; he should, instead, remove his foot from the stirrup and his right hand from the cantle and, staying near the horse's left shoulder, follow his movements forward, backward, or sideward until he eventually stands still. The rider then renews his efforts to mount.

d. After sitting down lightly in the saddle the rider puts his right foot in the stirrup, assisted by the right hand. He then holds the reins lightly in the left hand and strokes the horse on the crest and shoulders with the object of causing him to relax and stand still. The first mounting lesson should terminate without moving the horse forward.

e. As soon as the horse stands relaxed and reasonably calm with the rider on his back, the assistant takes the position of stand to horse, with his right hand holding the cheek piece, and causes the horse to move forward and change direction to the left. The horse is more easily moved forward if he is swung to the left at the same time. After thus moving a few steps the horse should be halted and petted; then the forward movement and turn to the left are repeated. The assistant then leads the horse for a considerable distance in a straight line before halting and petting him. When the horse leads freely and is relaxed and quiet the rider dismounts.

f. In dismounting it is very important that the horse is not frightened. Accordingly, assisted by the dismounted assistant, the rider dismounts quietly and carefully, endeavoring always to cause the horse to stand perfectly still.

g. If a horse is very difficult to mount, the rider should not be permitted to resort to violent methods. Such a horse should be returned to the exercise track and also given more dismounted leading and gentling before he is mounted.

h. (1) The first lesson of the second day's work should now be terminated by requiring the remounts to be led about dismounted for a few minutes and then lined up preparatory to falling out and returning to the stables.

(2) The second lesson of the second day's work should be a repetition of the above lesson, except that the riders, after mounting, remain a longer time in the saddle and cause the remount to move farther forward on straight lines than was

required during the first lesson. The assistant still remains near the head of the remount and renders the rider any necessary assistance. The remount is not allowed to move out of the walk while being led about.

67. Use of the aids.—*a.* During the first period the voice should be used freely to keep the remount calm and prepare him for the application of the aids. The use of the aids should be confined to three well-defined actions, which, in turn, should produce three simple effects. These actions and effects are—

(1) Action of the legs to cause the horse to move forward, to increase the pace, or to change to a faster gait.

(2) Action of the direct reins to cause the horse to change to a slower gait, to decrease the pace, or to halt.

(3) Action of the leading rein to cause the horse to change direction or to turn to the right or left about.

b. Once the remount is in motion, the reins should be used solely to maintain direction and to regulate the speed of the gait; no attempt should be made to set the horse's head and neck in position or to relax the jaw or poll.

c. The remount should not be halted abruptly; to slow up or halt, the rider should use the direct reins intermittently until the horse obeys. In slowing up or halting the remount, the rider's legs should not be used.

d. All changes of direction should be made on the arc of a large circle without insisting on an exact correctness or accuracy of movement. In turning, the leading rein should be used exclusively; the opposite rein may be used lightly to regulate the effect of the leading rein.

e. The legs should be applied by moderately increased pressure of the calves against the horse's sides; if this is insufficient, then by energetic, successive blows with the legs until the horse responds. This action of the legs may be supplemented by clucks of the tongue or light touches of the riding whip or a switch.

68. To teach submission to guidance and control.—*a.* Prior to requiring the following movements from the remounts while being ridden outside the riding hall, the movements listed below should be taught the remounts in the riding hall:

(1) To move forward; to halt.

(2) To turn to the right flank and to the left flank.

(3) To pass from the walk to the trot and from the trot to the walk.

(4) To turn to the right about and to the left about.

(5) To pass from the trot to the gallop and from the gallop to the trot.

b. In teaching remounts the above movements the officer in charge should, in general, be guided by the following procedure:

(1) The riders take the track at the walk, in rear of the man riding the most manageable remount and without regard to distances. By appropriate commands the riders are required to halt, move forward again at the walk, halt, and so on repeatedly.

(2) The riders, still moving at the walk, are next required to turn out of the column individually (when called by name) and halt on the middle line of the riding hall; then move forward and regain a place in the column.

(3) The riders are next required to pass from the walk to the trot and from the trot to the walk repeatedly.

(4) Next the riders are required to execute the about by tracing in column the course of a wide serpentine, first at the walk, then at the trot.

(5) The riders, being on the track at the walk, are required individually to pass from the front to the rear of the column; first, turning out of the column; next, taking the trot and moving on the track to the tail of the column, where they again take the walk.

(6) Finally, the riders are required individually to pass from the trot to the gallop and from the gallop to the trot repeatedly. In passing from the trot to the gallop, advantage should be taken of the corners of the hall to break the equilibrium of the horse so that he will take the gallop freely and correctly.

c. During the work outlined above there should be frequent short rests during which the rider dismounts and gentles his remount by petting, rubbing his legs, picking up his feet, etc. The remount is habitually required to stand still while the rider mounts and dismounts.

SECTION III

SECOND PERIOD

(Second 2 weeks)

69. Program of daily training.—*a*. During the second period the use of the exercise track is largely discontinued except for practicing the remounts in jumping. In addition to the rail jumps across the track, which during this period should reach a height of about 2½ feet, a shallow ditch about 2 feet wide should be dug across the track and gradually widened until at the end of the second period the remounts have been practiced in jumping a ditch about 4 feet wide.

b. The remounts should receive two training lessons per day, one in the morning and the other in the afternoon. The length of each lesson period should be about 2 hours. In general, the first 45 minutes of each lesson should be used for riding-hall instruction. Any remaining time should be spent in leading, gentling, picking up feet, grazing, riding on the road, cross-country riding, or drill.

c. After each lesson the remounts are thoroughly groomed and their feet handled.

d. While riding across country or on roads the remounts are frequently practiced in close and extended order drill formations. While riding across country the formations assumed should generally be such that each remount is well separated from the others. Thus the remounts are accustomed to going freely in any direction when alone.

e. While engaged in riding hall work the men should generally be required to ride without fixed distances.

f. The remounts are equipped with single snaffle bridles and stripped saddles. If the training period is materially reduced, they should begin to carry full field equipment during this period, the articles of equipment being added one at a time.

g. Each day's training should include a review of previous training prior to taking up any new exercises.

h. During this period the remount should become accustomed to eating from the feed bag. He should also become accustomed to having the pistol drawn, returned, and handled (including pointing the pistol) while at the walk. To start his training in becoming accustomed to firing and the sight of pistol targets, blanks may be fired in the vicinity of the exercise track during exercise and targets may be set up along the track. A practical and successful method of accustoming remounts to the sound of firing is as follows: all horses are on a large circle, the instructor with pistol and blanks on the inside; three or four assistants, each with a bucket of oats, are also on the inside with the instructor; as the pistol is fired all horses are turned to the inside of the circle for a bite of oats. After this exercise is given two or three times, the remounts will automatically turn toward the firing and come to the center as soon as the shot is fired.

i. The exercises outlined in the following paragraphs constitute the subject matter from which the officer in charge should make up his program of daily training.

70. Conditioning.—*a.* Conditioning is continued along the same general lines as during the first period. The object is to fit the remount eventually for field service. This service imposes two major requirements upon the horse: first, that he is able to march long distances day after day; second, that he is able to carry, while marching, the weight of the rider and his equipment.

b. (1) To be fit for marching, the horse must have three essentials: first, good feet; second, good legs; and third, good flesh.

(2) Good feet result from close attention to shoeing and trimming, their inspection each day, and the maintenance of clean, level standings.

(3) Leg injuries usually result from fast work, particularly on hard or uneven ground. A too hurried training is hard on joints and tendons. Many remounts are prone to interfere, especially while becoming accustomed to carrying weight. During this period it is important to watch for signs of interference on the horse's legs and protect the injured places by proper bandaging or boots.

(4) Good flesh is secured and retained by long, daily periods of slow work. During this work the walk is the gait principally used. In addition to long periods of slow work it is important to give close attention to the feeding, watering, groom-

ing, and bedding. Every effort should be made to build up the flesh of the thin remounts by increasing their grain and hay ration, by providing them with plenty of good drinking water, and by increasing the number of grain feeds per day.

c. To fit the remount for carrying his heavy load in the field it is necessary that the flesh and skin of his back be seasoned and toughened to withstand heavy pressure, heat, and sweat. The back muscles must be developed and strengthened for their weight-carrying job. To accomplish these results, the load to be carried by the remount should be gradually increased until he is capable of carrying a rider and full field equipment.

71. To improve carriage and action of the remount at the walk, the trot, and the gallop.—*a.* During the first period the remount is allowed to stretch out his head and neck to the front, the rider being careful only to keep the reins lightly stretched. No attempt is made, by means of the reins and legs, to gather the horse or to cause him to assume a correct attitude while in movement.

b. With the beginning of the second period, the improvement of the carriage and action of the horse at the walk, the trot, and the gallop should be started. Very gradually he should be brought to carry the head and neck properly and to engage the hindquarters. This adds to his handiness and balance and leads to his easier acceptance of guidance and control. It is accomplished by shortening the reins very slightly and increasing mildly the steady pressure of the legs. The rider should, to a greater extent than previously, keep the hands still without being too insistent lest he injure the horse's mouth, cause him to get behind the bit, contract the bad habit of throwing up his head, or show restiveness. The tranquility of the horse must be strictly maintained.

c. During work in the hall the riders should take the track at the walk without regard to distance. If the horse is restive and will not walk he should be taken out of the column and worked on an inside track or on a large circle. It is very important that the remount be taught how to walk, since in field service the greater part of the horse's work is marching, and to march well he must be a good walker. Having walked at attention for a few minutes, remounts should be permitted to rest in route order; that is, while resting they should continue at the walk, but should be encouraged to stretch out their heads and necks to the front, to relax completely, and rest. While thus resting, each rider should stroke his horse on the neck and crest and pat him on the shoulders and croup.

d. Upon resuming the walk at attention, the rider, in gathering his horse, should do so decisively but without brusqueness. The horse in movement should point his ears forward. To point his ears forward means that the horse is tranquil and is calmly considering his direction and the nature of the ground over which he travels. To point his ears backward indicates restiveness; the horse's thoughts are being too much attracted to the position or action of the rider. The rider must apply the aids with tact; he must do so in a manner to cause the horse to extend his stride at the walk and gradually to improve his carriage and action.

e. The exercise at the trot and gallop is executed in the same general manner. If the riders are sufficiently proficient and the remounts are going quietly, it is advisable to permit the horse to relax frequently and be ridden on a loose rein at the trot and gallop. The rider strokes his horse on the neck and crest, encouraging him to relax and move freely. When possible at this stage of training, all galloping should be done in a direction away from stables and preferably up a gentle slope.

72. To train remount to submit willingly and easily to guidance and control.—*a.* During the second period the training prescribed for the first period (par. 68) should be continued so as to increase the handiness and willingness of the remount while executing forward movements, halts, changes of gait, and changes of direction. Through much practice of these exercises the horse gradually learns to engage his haunches properly, develops and strengthens his muscles, and acquires a correct carriage of his head and neck. In this manner the horse slowly acquires a true balance, free from bad habits and unnatural posture.

b. All training in starting, stopping, and turning should follow a gradual and well-ordered progression; carried to excess, such training may injure the horse's joints or tendons and destroy his tranquillity.

c. The basis of all training outlined in this paragraph is freedom in the forward movement.

73. To accustom remount to use of arms.—*a.* During the second period the pistol should be drawn by the rider while mounted and handled at the various gaits.

b. To accustom the remounts to the pistol being handled, the men first practice drawing, pointing, and returning the pistol while riding at the various gaits; next the remounts are ridden quietly on the mounted pistol course while the riders simulate fire on the targets; finally the horses are ridden on the mounted pistol course while the riders fire a few shots to the right and left, keeping the pistol well away from the horse's head. If caliber .22 pistols are available it is well to start with them in training the horse, since the noise and shock are not so great as they are with the caliber .45 pistol. The horses should be ridden on a loose rein or with fingers completely relaxed so that on throwing the head they will not receive at the same time a bump on the mouth.

Section IV

THIRD PERIOD

(Third 2 weeks)

74. Program of daily training.—*a.* During the third period, the first 45 minutes of each day's training should be conducted in a riding hall. Any remaining time (which ordinarily should be from 45 minutes to 2¼ hours) should be spent out of doors.

b. The remounts are equipped with the double snaffle bridle, if available, otherwise with the single snaffle bridle. Except when training the remounts to jump obstacles, the men ride with spurs. They may carry a riding whip or switch, if necessary.

c. Certain exercises prescribed in the first and second periods should be constantly repeated and practiced during the third period. The more important of these are to mount and dismount (the horse standing still and tranquil); to rest the horse in movement; to improve the carriage and action of the horse at the walk, trot, and gallop; to train the horse to submit willingly and easily to guidance and control; and to accustom the new horse to the use of arms and the pressure of ranks.

d. The exercises outlined in the following paragraphs constitute the subject matter from which the officer in charge should make up his program of daily training.

75. Conditioning.—*a.* Conditioning is continued along the same general lines as during the second period. The remount's physical condition is improved by continued slow work and unremitting attention to his feeding, watering, bedding, grooming, shoeing, and clipping. In addition the remounts should be taken on short practice marches every second or third day, starting with a march of 5 miles and gradually increasing the distance to approximately 15 miles.

b. The outdoor exercise has two principal objects in view. One is the gradual development and strengthening of the muscular system of the horse. This is primarily secured through the long periods at the walk, broken occasionally by short periods at the trot. The other is the general development of the respiratory system of the horse. This is chiefly secured through the longer periods at the gallop. While it is true that any gait tends to the development of both the muscles and the respiration, the walk is regarded primarily as the gait for muscle building, and the trot and gallop as the gaits for improving the respiratory system.

c. The time to be devoted each day to the walk, the trot, and the gallop, respectively, cannot be prescribed exactly. It is dependent upon many considerations, such as the actual state of condition of the horses, their health, weight, appearance, and animation. Even the weather must be taken into consideration. These are all matters in connection with which the officer in charge must be very observant and about which he must exercise good judgment.

d. Training in the jumping of obstacles and in passing over rough and somewhat difficult ground should be included during the course of each conditioning ride.

e. At the weekly inspection of remounts, each horse should be examined minutely as to the progress of his conditioning. There should be evident a marked improvement in health and strength. The eyes should be wide-awake, the coat glossy, the skin soft and supple, and the muscles firm. The flanks should not be tucked up; they should be full and round and muscular. The legs should be free from bursal enlargements or bony growths, the tendons cool and sound, the joints showing no indication of sprains. The muscles of the legs should be firm; as conditioning progresses, the muscles of the legs and of the entire body should become firmer and stand out more and more.

76. Manner of applying the aids.—*a.* In all movements involving the guidance and control of the horse the rider should now, at the commencement of the third period, enter upon the use of the aids in combination. Heretofore, due to greenness and awkwardness of the new horse, the aids have been applied in the simplest possible manner.

b. Hereafter to start the horse, to change to a faster gait, or to increase the speed of a gait, the legs and the weight should be used in combination; conversely, to stop the horse, to change to a slower gait, or to decrease the speed of a gait, the reins and weight should be used in combination; while to turn to the right, to the left, or about, the effect of the leading rein should be more and more reinforced by the effect of the bearing rein until, near the close of this period, it should be possible to guide the remount by means of the bearing rein alone in combination with the effect of the legs and weight.

c. Care and exact use of the aids at the walk in the riding hall, in the execution of the figures and exercises prescribed, are essential; they are the basis of the training as a result of which the rider, out of doors, at any gait or speed is able to guide and control his horse. The gradual substitution of the bearing rein for the leading rein, even though the rider holds the reins in both hands, is of the greatest importance. The remount is thus prepared for his eventual use in the ranks of the troop, when the trooper habitually rides with the reins in one hand.

77. To improve handiness and balance of remount.—*a.* All the exercises outlined in this paragraph are fully described in chapter 1. The application of these exercises to the training of remounts will be easily understood if in reading that chapter it is kept in mind that the horse is now being trained instead of the rider.

b. In starting, stopping, backing, and in changing the gait, the horse must know how and be willing to engage his haunches; he must also have the appropriate muscular development in order to do so effectively. He must in addition be light in the fore hand. Being well balanced, he is then the more easily controlled. These results are obtained through the daily practice of such exercises as will train the horse in obedience and develop him physically. These exercises are—

(1) Being at the halt, to move forward, to halt, to move backward, and again to move forward, repeatedly.

(2) Being at the trot, to take the slow trot, the walk, the slow trot, and again the trot, repeatedly.

(3) Being at the gallop, to take the trot, the walk, the trot, and again the gallop, repeatedly.

(4) Being at the gallop, to increase and decrease the pace.

c. The horse must not only be well-balanced and obedient in the manner described in *b* above, he must also know how and be willing to change direction to the right, the left, or about. These results are best attained by the use of certain riding hall figures and exercises, the energetic execution of which constantly develops the horse's muscular strength, sense of balance, and handiness. In the training of horses it should not be forgotten that exactness in the execution of the figures and exercises has chiefly to do with obedience; that the carriage and manner of the horse, his vigorous and spirited action, have chiefly to do with physical culture. The figures and exercises referred to are—

(1) Serpentines.

(2) Circling individually.

(3) The half turn in reverse.

(4) The figure of 8.

(5) The about.

(6) The gallop departs.

(7) Work by threes.

d. In addition to the above, the horse must be so obedient to the aids as to be easily moved in any direction, at any gait, and in the manner desired by the rider. These results are to be attained by practice of the following exercises:

(1) To leave ranks.

(2) To increase and decrease the pace.

(3) To gait.

(4) To practice, individually, the charge at full speed.

78. To train remount to jump obstacles.—*a.* The dimensions of obstacles should be increased in proportion to the progress of the horse. The eventual limit for high jumps should be 2½ feet and for broad jumps 5 feet. Obstacles without wings should have ample width of front so as not to tempt the remount to avoid jumping by passing around the obstacle.

b. The training of remounts to jump should be conducted in the same general manner as prescribed in chapter 1 for teaching recruits to jump.

c. In teaching the untrained horse to jump he should first be practiced over small obstacles at the walk and trot without the rider on his back. After considerable practice in this method of jumping, the same training is repeated with the rider on his back. While receiving this training the remount should not be sent rushing over the obstacles. He should, on the other hand, be allowed to approach the obstacle calmly and learn to jump by exertion rather than by a flying leap. When trained in this manner the remount develops the muscles of his hindquarters, loins, back, and shoulders; learns how to use his head and neck as a balance; and acquires sureness in handling his feet. Having completed the above training, the remount is ready for training in taking the jumps at a gallop. In all preliminary training in mounted jumping, a take-off jump, placed on the ground or a few inches above the ground and about 2 to 2½ feet in front of the jump, should be used. This is an automatic aid in helping the remount measure his stride and adjust his take-off. It also serves to prevent rushing in his early training.

d. In conformity with the training scheme outlined above, the remounts should continue during the third period to receive daily training in jumping low obstacles on the herding track without the riders on their backs.

e. In addition to the training in jumping without the riders on their backs the remounts are given a little training each day in jumping low obstacles with the riders up. For this training the jumps are kept at least 6 inches lower than the jumps without the riders on their backs.

f. In approaching an obstacle at the walk or trot the horse should be permitted to increase his last few strides in order to acquire a little spring and momentum. In jumping small obstacles at the gallop, the speed of the gait should be moderate and the horse should jump in his galloping stride without increasing the speed before or after taking the jump.

79. To accustom remount to packed saddle and to submit to conditions of field service.—*a.* During the first and second periods of training, the remounts are gradually accustomed to the presence on their backs of the various items of full-pack field equipment. During the third and fourth periods of training, the remounts should be required to carry this equipment during all their practice marches in order that the weight-carrying muscles of their backs may become stronger, and the skin and flesh may become somewhat seasoned to the pressure, friction, heat, and sweat caused by the fully packed saddle.

b. During marches across country and on the road the riders should be frequently required to dismount and lead. The horses should be trained to lead forward at a brisk walk at the rate of approximately 4 miles per hour. Those remounts that are inclined to lag back should be urged to walk faster by means of light touches on the hindquarters from a switch carried in the rider's left hand. Some practice should also be given over low obstacles and ditches.

c. In order that the remounts may become accustomed to eating their grain from the feed bag, they should be fed once a day for 2 or 3 days in this manner. Care and patience must be used when first putting on the bags so as to avoid frightening the horses.

d. While marching in any formation, remounts should be accustomed to leaving ranks. This training may be accomplished by directing individual riders to leave the formation, ride to a designated place, and then rejoin.

SECTION V

FOURTH PERIOD

(Fourth 2 weeks)

80. Program of daily training.—*a.* During the fourth period, the first 45 minutes of each day's training should be conducted in a riding hall. Any remaining time should be spent out of doors. On days when a long practice march is scheduled the entire time should be spent out of doors.

b. The remounts are equipped with the curb bit and snaffle. The troopers ride with spurs; with particularly restive horses blunt spurs are worn. For a horse that especially requires it, the double snaffle bridle may still be employed.

c. All the exercises and work of the previous periods should now be repeated in connection with the use of the bit and bridoon (curb and snaffle bits). This repetition and review have three principal objects in view. The first is to carry to completion the training of the horse in obedience to the aids. The second is, by means of long conditioning and cross-country rides, to render the horse a good goer out of doors at all gaits. The third is to accustom the remount to the use by the rider of the curb bit and snaffle, the reins being held in one hand.

81. Adjustment of bit and bridoon.—When first used on a remount the curb bit and snaffle should be most carefully adjusted in a manner to protect the mouth of the remount from injury. With this in view both bits should be placed high in the mouth and the curb chain made quite loose. The rider, moreover, should use the snaffle bit almost exclusively. The curb bit should be brought into play very lightly, gently, and gradually. After a few days the bits and the curb chain may be adjusted more nearly correct; if all goes well they may, after a day or two more, be properly adjusted as described in paragraph 20.

82. To accustom remount to effect of bit and bridoon.—During the first few days' work with the curb bit and snaffle it is better to go out of doors. Any work indoors should be mild in character, nothing of an exacting nature being demanded of the horse until he has fully accepted and goes well into the bit and bridoon. There is now great danger that the horse may "get behind the bit." This must be most carefully guarded against by the officer in charge and by each rider. In all exercise and work the reins should be held lightly taut and quite long, a lighter pressure on the curb than on the snaffle. The horse should be encouraged, by tactful use of the legs, to extend his head and neck to the front and to maintain light contact with the bit. During periods of rest, whether standing still or moving, the reins should be allowed to swing loose, thus according the horse full liberty to relax.

83. Swimming.—When facilities are available, the remounts should, during this period, receive training in swimming. They should first be trained in fording streams, and any fear of water eliminated before requiring them to swim. For the first lesson in swimming they should be allowed to swim at liberty, that is, following older horses that swim freely and willingly or being led behind a boat. When they have learned to swim freely and willingly, the riders should swim with their horses as prescribed in paragraph 50.

CHAPTER 3

ANIMAL MANAGEMENT

SECTION I

HORSE AND MULE

84. General structure.—The body of the horse, like that of man, is made up of a skeletal system, a muscular system, a digestive system, a respiratory system, a circulatory system, a nervous system, a urinary system, a reproductive system, and an outer body covering of skin and hair.

a. Skeletal system.—The skeleton of the horse is made up of about 205 bones. Bones give the body form and rigidity and form cavities for the protection of vital organs. The bones and their joints together form a system of levers and pulleys which through muscular action gives the body power of movement. Joints are of special importance, for they are frequently injured and are the seat of many unsoundnesses.

b. Muscular system.—The muscles or red flesh make up most of the bulk of the body, and by their action produce all movements. The body of the muscle is usually attached to some bone while the muscle is continued as a fibrous band called a tendon, which may pass one or more joints and attach to some other bone. Muscles are seldom injured, but the tendon of the muscle is more liable to injury.

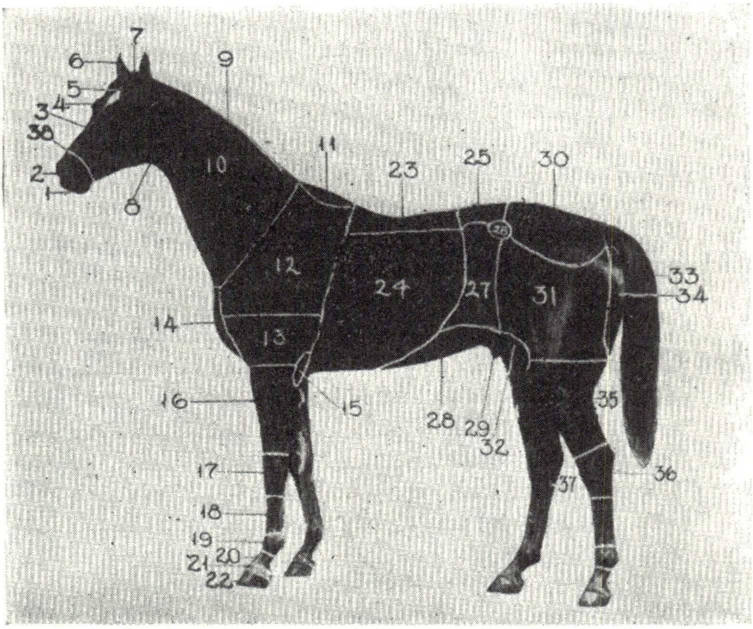

FIGURE 11.—Regions of the horse.

1. Lips
2. Nostril
3. Face
4. Eye
5. Forehead
6. Ear
7. Poll
8. Throat
9. Crest of neck and mane
10. Neck
11. Withers
12. Shoulder
13. Arm
14. Breast
15. Elbow
16. Forearm
17. Knee joint
18. Cannon
19. Fetlock joint
20. Pastern
21. Coronary band
22. Hoof
23. Back
24. Ribs or costal region
25. Loins
26. Point of the hip
27. Flank
28. Abdomen or belly
29. Sheath
30. Croup
31. Thigh
32. Stifle
33. Tail
34. Buttocks
35. Leg
36. Hock
37. Chestnut
38. Muzzle (includes lips, mouth, nostrils, and nose)

c. Digestive system (par. 91*a*).—The digestive system is really a long tube beginning at the mouth and ending at the anus. Its function is to receive and digest the food. The digested part of the food passes into the blood stream as food for the muscles, and the undigested portion is passed as dung. As the horse naturally eats grasses and not grains, he must be supplied with bulky food. The droppings should be formed in pellets and flatten slightly on striking the ground.

d. Respiratory system.—The organs of the respiratory system are the nasal passage, the pharynx, the larynx or voice box, trachea or windpipe, the bronchi, and the lungs. The purpose of this system is to supply the blood with oxygen and to throw off the waste gases. Breathing should be free, soft, and noiseless. After fast work the breathing is heard as a rushing sound of air but there should be no harsh, fluttering, whistling, or roaring sound. After cessation of exercise the breathing rate should subside quickly to normal. Cinching too tightly or placing the cinch too far to the rear will interfere with normal breathing.

e. Circulatory system.—The heart pumps the blood out through the arteries and small capillaries to all parts of the body. The blood returns to the heart through the veins. The blood passes through the lungs where it gives up its waste gases and receives its supply of oxygen. The blood circulating in the walls of the stomach and intestines absorbs the digested food and makes it available for the tissues of the body. The arteries, as a rule, lie deep in the tissues and are well protected from injury. When the horse is at rest the pulse rate should be 36 to 40 per minute. The rate increases with exercise. The rate will be about 60 to 70 after a 5-minute gallop.

f. Nervous system.—The nervous system consists of the brain, the spinal cord, and the nerves running to all parts of the body. The nerves are of two kinds; those which carry messages from the brain and those that carry sensations to the brain. Nerves usually lie close to arteries in their course.

g. Urinary system.—The function of the urinary system is to remove from the body waste products extracted from the blood by the kidneys. The horse passes urine several times daily in quantities of a quart or more, 3 to 6 quarts being passed in 24 hours. During the act of urination the horse spreads and extends the hind legs to the rear and stands on the toes of the hind feet, grunts, and assumes a very awkward position.

h. Reproductive system.—The function of this system is the bearing of young. The stallion will breed at all seasons of the year but the usual active season of the mare is from January to July, when the mare usually comes in "heat" for a period of 3 to 7 days every 18 to 21 days. During the "heat" period the mare is nervous, restless, urinates frequently and, during this period, trains or works less satisfactorily. With reference to age and sex, horses are described as follows:

Foal—Either sex under 1 year of age.

Yearling—Over 1 year but less than 2 years of age.

Filly—A female horse under 4 years of age. *Example:* Filly foal, yearling filly.

Mare—A female over 4 years.

Gelding—A male horse which has been castrated.

Stallion or horse—A male horse over 4 years.

Colt—An unaltered male under 4 years.

Ridgling—A male with but one testicle removed, the other being retained in the belly.

i. Skin and hair.—The skin forms the outer protective covering of the body, throws off waste products through sweating, assists in regulating the body temperature, and produces the hair of the body coat and the horn of the hoof, chestnut, and ergot. The ergot is the small horny growth on the back of the fetlock joint. The base of the ears is the first place in the horse where sweating takes place; the neck, the flank, the side of the chest, the back, and the croup follow in order. All hair except the mane and tail is shed in the fall and in the spring. During shedding, especially spring shedding, the vitality and condition of the animals are somewhat lowered and the skin is more susceptible to infections.

85. Temperature.—The normal body temperature of the horse at rest is about 100° F., but may vary 1° in either direction. The temperature varies with exercise, excitement, and air temperature. Hard, fast, or prolonged work under a hot sun may build the body heat up to 104° to 107° F. When the temperature reaches this reading, the horse is approaching "overheat." The temperature is taken with a clinical thermometer in the rectum.

86. Posture.—The standing posture is the most common posture of the horse. Normally both front feet will be on the same transverse line and bear weight equally. Any other posture of a forefoot is spoken of as "pointing" and is an indication of trouble. The hind limbs are rested alternately; rarely does the horse stand with both hind feet squarely together on the ground. The horse can maintain the standing position without tiring and can sleep standing. The horse lies down either obliquely on the chest with the legs somewhat folded under the body and head extended with the chin or teeth on the floor, or flat on his side with the legs and head extended.

87. Age.—The average duration of life of the horse is about 20 years. Horses are capable of reproducing at about 1½ years of age but they are not bodily mature until about 6 years of age. The period of usefulness will average about 10 or 11 years. Horses 10 years of age or over are better able to stand the hardships of campaign than younger horses. The age of a horse may be estimated by the appearance of the front teeth by those experienced in such judgment. The age of all military animals may be determined by examining the horse record card, W.D., Q.M.C. Form No. 125, where the foaling date is recorded.

88. Measurement.—*a.* The *height* is the vertical distance from the ground to the highest point of the withers and is expressed in hands and inches, a hand being 4 inches.

b. The *girth measurement* is the circumference of the body in inches measured at the juncture of the withers and back. Size in girth is desirable as it indicates heart and lung space. A well conformed horse 16 hands in height should have a girth measurement of about 78 inches.

c. The *bone measurement* is the circumference in inches of the fore cannon midway between the knee and fetlock. This measurement will vary between about 7¼ and 8½ inches in riding horses standing 15 to 16 hands in height.

89. Gaits.—The walk, trot, and gallop are the only gaits desired for military animals.

a. The *walk* is marked by four beats. The feet are raised successively and planted in the order in which raised; for example, right fore, left hind, left fore, right hind. The regulation walk in marching is 117 yards per minute or 4 miles per hour.

b. The *trot* is marked by two beats and a period of suspension. The horse springs from one diagonally disposed pair of feet to the other; between the beats all feet are in the air. The right fore and left hind are called the right diagonal, and the opposite diagonal is the left diagonal. The *regulation marching* trot for cavalry is 9 miles per hour, for artillery 8 miles per hour. The *slow trot*, used for equitation instruction, is 6 miles per hour.

c. The *gallop* is marked by three beats and a period of suspension. For example, the horse is galloping with the right lead. The first beat is marked by the left hind foot striking the ground, the second by the nearly simultaneous planting of the right hind and left fore, and the third by the planting of the right fore. The regulation *gallop* is at a rate of 12 miles per hour. The *extended gallop* is at a rate of 16 miles per hour.

90. The mule.—The mule results from the mating of a jack (male donkey) with a mare. Mules are suited to slow draft and packing and require less food than horses. The above paragraphs apply to mules as well as to horses.

Section II

FEEDING AND WATERING

91. General.—The health and condition of animals depend largely upon the kind of food received and the hours of feeding. To feed so as to obtain the best results requires a knowledge of the digestive system of animals, the food elements required for their sustenance, the correct proportions of these elements, the proper methods of preparation, and the best hours of feeding.

a. The organs of the digestive system form the alimentary canal and consist of the mouth, pharynx, esophagus, stomach, small intestines, and large intestines, consisting of the caecum, large colon, small colon, rectum, and anus. The large glands, such as the salivary glands, liver, and pancreas, are accessory organs.

b. The stomach is the principal organ of digestion. Its capacity is from 3 to 4 gallons. The intestines are also organs of digestion and have a much greater capacity than the stomach.

c. Food is taken into the mouth and masticated. It passes through the pharynx and esophagus into the stomach, where it is partially digested and prepared for complete digestion in the intestines. In the intestine, digestion is completed and the nutriment and moisture extracted. The residue is discharged through the anus as dung.

d. The time required for stomach digestion varies with the class of food. Grain requires approximately twice as long as hay. For this reason hay should precede the grain ration or else follow it by an hour or two. The latter is best when a large hay feed is given, as in the evening. A small feed of loose hay shortly before the grain takes the edge off the appetite and prevents bolting. It greatly stimulates the saliva and stomach secretions and is an aid in digestion by furnishing the bulk necessary for proper stomach digestion.

e. Water remains in the stomach for only a short time, passing quickly into the intestines, where it is collected in a part of the large intestines called the caecum. If given soon after feeding, it flushes a considerable portion of the food from the stomach and into the intestine before it is digested. In general, watering should be done before feeding or not until an hour or two afterward. Animals drink very little or none if watered very early in the morning and for that reason the morning water is best given shortly before the drill period or marching.

92. Food constitutents.—*a.* Food constituents are classified according to the function they perform in the animal system as follows:

(1) Flesh making; such as linseed meal, oats, peas, beans, bran, etc.

(2) Fat, heat, and energy producing; such as corn, hay, oats, etc.

(3) Bone making; such as grass, hay, bran, bonemeal, milk, etc.

(4) Bulk supplying, necessary to digestion (fibrous and woody material); such as hay, grass, bran, etc.

(5) Watery; such as carrots, potatoes, green grass, etc.

b. There is no one food containing all of the required constituents, hence two or more must be combined to obtain the desired ration. Natural grasses come nearer to serving this purpose than any other food.

93. Proportions of various food components in a ration.—The components of any ration depend upon the class and condition of the animals, the work required of them, the variety of available foods, the kind of shelter afforded, the climate, the season, etc. The food must be selected in such proportions as to form a balanced ration and fed in such quantities at such hours as to enable the animal to obtain the greatest possible amount of nutrition from them. The forage ration with its permissible substitutes as given in paragraph 106 furnishes a ration which is sufficiently balanced for service use.

94. Oats.—*a.* Oats are the safest of all grains for animals and are standard for issue in the Army. All other grains are regarded as substitutes.

b. Oats may be safely fed in quantities up to 16 pounds per day, provided the animal is performing heavy work. However, because of the small size of their stomachs, they should be fed no more than 4½ pounds at one feeding. Under ordinary conditions the amounts given in paragraph 106 are sufficient. Horses and mules, when rested for a length of time, should be given no more than 6 pounds daily.

95. Corn.—*a.* Corn is less desirable than oats by reason of its tendency to produce heat and fat. It is best fed during the colder months of the year and combined with oats and hay, as indicated in paragraph 106, it forms a well-balanced ration.

b. Corn should, when practicable, be fed on the cob. One feed consists of from 6 to 12 ears, depending on the size of the ears, the amount of work performed, and the individual animal. When necessary to feed shelled corn, mix with dry bran or oats, or both, to prevent bolting.

c. Change from oats to corn should be made gradually, substituting about 2 pounds of corn for 2 pounds of oats each successive week.

96. Barley.—*a.* Barley is a very good food for horses and may be safely substituted for oats and fed in the same quantities. The change from oats to barley should be gradual and extend over a period of at least 2 weeks.

b. Barley may be fed whole, but, due to its excessive hardness, it is usually crushed or soaked in water for 2 or 3 hours before feeding.

97. Rye.—Rye is not regarded as a good forage. Should its use become necessary, it should be fed in small amounts and mixed with other feed such as oats or bran. Whenever possible, rye should be rolled or crushed before feeding. In emergency it may be mixed, in amounts of 6 to 10 pounds, with some bulkier grain such as bran or oats.

98. Wheat.—Wheat alone is not a safe food for horses. Only 1 or 2 pounds should be fed at first and the amount gradually increased to a maximum of 6 pounds a day. Wheat should be combined with some bulky grain or mixed with chaff or chopped hay. It is desirable to roll wheat before feeding.

99. Rice.—Rough rice, when rolled, crushed, or coarsely ground, may be fed in quantities up to about one-half of the grain ration. In emergency it may be fed in quantities up to 14 pounds daily.

100. Linseed meal.—Linseed meal is an excellent food for animals that are in a run-down condition. Its effect is slightly laxative and its use will improve the gloss of the coat. It is a useful feed during the spring shedding. As linseed meal is unpalatable to most horses, it should be mixed with other feed, such as oats and bran. As a general rule, it should not be fed in larger amounts than one-half pound daily; smaller amounts such as a heaping mess-kit spoonful three times a day mixed with grain often are sufficient.

101. Bran.—*a*. Bran is an excellent food for animals. As an adjunct of a ration consisting largely of grain, it is most useful on account of its mild, laxative effect. It helps in building bone and muscle without tending to flatten and adds to the general tone and condition of animals.

b. If dry bran is fed daily, mixed with oats or other grain, about 2 pounds is the proper amount. Feeding bran produces more thorough mastication of the grain, adds bulk to the grain ration, and supplies slightly laxative and tonic properties. The quantities should be reduced when animals are grazed or fed alfalfa or other foods that have a laxative action.

c. An occasional feed of bran in the form of a mash is relished by most horses and it is quite beneficial. In this form bran has a more pronounced laxative action. Bran mashes are best given the night before a day of idleness or light exercise.

d. A bran mash is made by placing 2 or 3 pounds of bran in a feed pan, sprinkling a mess-kit spoonful of salt over the top, and then adding boiling water while stirring until the mixture is well-dampened but not sloppy. Cover or wrap the container in a horse cover and allow it to stand until cool enough not to burn the mouth.

102. Salt.—Horses show great fondness for salt and thrive best when regularly supplied with it. It should always be available in the form of cake or rock salt placed in or conveniently near the manger. Feed pans for grain should not be used as salt containers. Under field conditions animals should be given salt about twice a week in the form of rock salt, one heaping mess-kit spoonful mixed with the grain in the feed bag.

103. Hay.—*a.* Hay cannot be wholly dispensed with for any considerable time unless it is replaced by a similar article, such as straw, as bulk is necessary for the proper performance of digestion. Oat straw is the best substitute for hay. The use of barley or rye straw is not recommended. Under field conditions, when issues of hay are diminished or entirely suspended, it is important that animals be grazed or fed such roughage as can be gathered.

b. All classes of hay, with the exception of alfalfa, are of practically equal feeding value.

c. Clover hay makes a good forage for draft animals. It has a laxative effect and when fed alone may make animals "washy." Clover frequently appears mixed with timothy, and when it does not comprise over 50 percent of the mixture it is an excellent hay for general use.

d. Alfalfa is quite laxative in its effect on animals unaccustomed to its use. In feeding properties alfalfa is more like the grains than like other hays, and combines especially well with a grain ration of corn. The change to a ration containing alfalfa should be made gradually beginning with 2 or 3 pounds daily. Alfalfa should not comprise much more than one-half of the hay allowance. As the most nutritious part of alfalfa is the leaves which shatter easily, great care should be exercised in feeding to see that this part is not wasted. Small amounts of alfalfa, 2 or 3 pounds, are most economically fed as "chop." Alfalfa is not a good feed for use during hot weather.

e. Chop is hay cut in short lengths. It is best fed mixed with grain and insures more thorough mastication of the grain. A pound or more of chop may be mixed with each grain feed. As chopped hay is likely to be dusty it is well to dampen it by sprinkling shortly before feeding.

f. Hay that is very dusty should be sprinkled about one-half hour before feeding. Sprinkling with salted or sweetened water will often tempt the appetite of shy feeders.

104. Grazing.—*a.* Grazing is very beneficial to the animal.

b. In turning animals out to graze for the first time, care must be taken that they do not overeat. Flatulent colic is almost sure to result from grazing on wet or frosted alfalfa or clover. When grazing is sufficient and regular, the feeding of bran should be discontinued.

c. In order to augment the possible scanty ration, animals in the field should be allowed to graze at every opportunity.

105. Compressed forage.—*a.* The compressed forage ration is primarily for field service. Present specifications cover three types of varying compositions of alfalfa meal and cane molasses with other ingredients. Two types may be used as substitutes for hay only, while the third type may be substituted for hay and grain.

b. Where possible the change from the usual ration to the compressed ration should be made gradually.

c. Special precautions must be taken to prevent this compressed ration from becoming damp as it molds very easily and, when fed in this condition, causes intestinal disorders and diarrhea.

106. Forage ration.—*a. Classes.*—(1) *Garrison ration.*—Issued at permanent or semipermanent camps.

(2) *Field ration.*—Used in the field, on marches and maneuvers.

b. Division of horses for rationing purposes.—(1) Small horses, issued to foreign stations.

(2) Light horses, 1,150 pounds in weight and less.

(3) Heavy horses, more than 1,150 pounds in weight.

c. Allowances; components.—The following tabulation shows the allowances prescribed in each class of ration for each division of horses mentioned in *b* above and for mules:

(1) *Garrison ration.*

	Small horses	Light horses	Heavy horses	Mules
	Pounds	*Pounds*	*Pounds*	*Pounds*
Grain	7	10	12½	8
Hay	14	14	15	14
Straw	5	5	5	5

(2) *Field ration.*

	Small horses	Light horses	Heavy horses	Mules
	Pounds	*Pounds*	*Pounds*	*Pounds*
Grain	9	12	14	10
Hay	14	14	16	14

d. Substitutes and equivalents.—(1) Not to exceed 3 pounds of bran may be substituted for a like weight of grain, or one-half pound of linseed meal for 1 pound of grain.

(2) In foreign possessions the substitution of palay, copra meal, or any native product for the grain ration, and native grasses for hay, is authorized where a saving can be effected.

(3) In the issue of forage, 10 pounds of hay is considered the equivalent of 15 pounds of corn fodder or grain sorghums.

e. Variations.—(1) The variations permissible in the garrison ration and which can be furnished in a practical manner are as follows:

For horses under 1,150 pounds

Standard ration	*Variation No. 1*	*Variation No. 2*
10 pounds oats	8 pounds oats	6 pounds oats
14 pounds hay other than alfalfa	5 pounds alfalfa hay	2 pounds corn
	9 pounds hay other than alfalfa	5 pounds alfalfa hay
		9 pounds hay other than alfalfa
—	—	—
24 pounds	22 pounds	22 pounds

For horses over 1,150 pounds

Standard ration	*Variation No. 1*	*Variation No. 2*
12½ pounds oats	10 pounds oats	6 pounds oats
15 pounds hay other than alfalfa	5 pounds alfalfa hay	3 pounds corn
	10 pounds hay other than alfalfa	6 pounds alfalfa hay
		9 pounds hay other than alfalfa
—	—	—
27½ pounds	25 pounds	24 pounds

For mules

Standard ration	*Variation No. 1*	*Variation No. 2*
8 pounds oats	5 pounds oats	3 pounds oats
14 pounds hay other than alfalfa	5 pounds alfalfa hay	2 pounds corn
	9 pounds hay other than alfalfa	6 pounds alfalfa hay
		8 pounds hay other than alfalfa
—	—	—
22 pounds	19 pounds	19 pounds

(2) The following variations are suitable when oats are not available:

For horses under 1,150 pounds

Variation No. 1	*Variation No. 2*
5 pounds crushed wheat	5 pounds crushed wheat
5 pounds rolled barley	5 pounds corn
6 pounds alfalfa hay	6 pounds alfalfa hay
5 pounds prairie hay	7 pounds timothy hay
—	—
21 pounds	23 pounds

For horses over 1,150 pounds

Variation No. 1	*Variation No. 2*
6 pounds crushed wheat	6 pounds crushed wheat
6 pounds rolled barley	6 pounds corn
6 pounds alfalfa hay	6 pounds alfalfa hay
5 pounds prairie hay	7 pounds timothy hay
—	—
23 pounds	25 pounds

For mules

Variation No. 1	*Variation No. 2*
3 pounds crushed wheat	3 pounds crushed wheat
3 pounds rolled barley	3 pounds corn
3 pounds alfalfa hay	3 pounds alfalfa hay
10 pounds prairie hay	10 pounds timothy hay
—	—
19 pounds	19 pounds

107. Food requirements.—*a.* When animals gnaw woodwork and eat dirt and dung, the cause may be a lack of mineral matter in the food or insufficiency of hay in the ration. In the field, when hay cannot be obtained, grazing should be resorted to at every opportunity. Failing grass, the animal should be permitted to consume any form of roughage, such as straw, bark, leaves, etc., with which his appetite can be tempted.

b. For idle animals that amount of food is sufficient which maintains the weight of the body satisfactorily. The average maintenance ration should be about 18 pounds of forage for each 1,000 pounds of body weight. It may consist of hay only or of about 4 pounds of oats and 14 pounds of hay. Mules and horses absolutely idle in corrals will retain their weight on a daily allowance of 18 pounds of good hay.

c. When the work of animals is increased during conditioning for field service, the grain should be increased as the condition of the animal indicates; the amount fed to some horses may be somewhat above the prescribed allowance. To permit of this increased feeding, unit commanders should accumulate ration savings during the season when work is lighter, such as in winter or during target season, and the animals are not being fed the full ration.

108. Feeding in garrison or camp.—*a.* Grain is fed about first call for reveille. This early morning feed should be small; 2 or 3 pounds per animal is ample. Just before saddling, harnessing, or packing, the animals may be watered. During the time the animals are out, feed boxes and hayracks should be thoroughly cleaned and 3 or 4 pounds of fresh hay put in the latter. After morning stables the animals should be watered and then tied in their stalls. At noon their second grain may be given. Assuming that 9 pounds is to be fed an animal, the division of about 2 for morning, 3 for noon, and 4 for evening is correct. With 12 pounds, 3 for morning, 4 for noon, and 5 for evening is correct. When the grain is more than 12 pounds it is better to feed four or five times than three.

b. Upon completion of the noon feed the animals are tied outside, except in very hot or inclement weather. After evening water call they are again tied in and should find the remainder of their hay ration in their hayracks. About supper call the evening feed of grain should be given.

c. The advantage of feeding in this way is that a small feed of grain is given in the morning, which the animal will thoroughly digest before going to work. Feeding the hay before the other grain feeds keeps the horses from being greedy and bolting their grain without masticating it.

d. In camps where feed boxes are not available feed bags should be used. Grain is fed by issuing the proper amount to each man who feeds his horse individually. Feed bags must be adjusted so that the animal can reach his grain easily; otherwise, tossing of the head and loss of grain will result. Any grain left in the feed bags should be spread on a paulin to dry and be used for the next feed. Under no circumstances should such grain be scattered on the ground in front of animals. To do so invites serious dirt or sand colic. Any grain spilled out of a feed bag by an animal should be swept up for similar reasons.

e. In the field, hay should be fed by breaking up the bales, distributing it along the picket line in small quantities, and frequently renewing it. A detail of men should be kept with the animals to prevent the hay from working back beneath the feet and being trod upon and wasted. To avoid undue waste, bales of hay should be carried to the picket line and broken only as actually needed.

109. Feeding on the march.—On the march the principle of feeding small amounts and often should be adopted. At hourly halts horses should be allowed to graze. When the duration of the march is over 5 or 6 hours, a midday halt of an hour or more should be made and animals given a feed of grain.

110. Feeding special cases.—*a. Animals that bolt their food.*—To overcome this habit a considerable amount of bran should form a part of each feed. If this does not produce the desired results, nail slats across the manger so as to divide it into compartments or place a few large stones in the feed box.

b. Windsuckers and cribbers.—If possible, feed windsuckers and cribbers apart from other animals.

c. Animals that "stock."—(1) This is a condition in which the legs swell in the fetlock region.

(2) The rations of animals predisposed to this condition must be carefully regulated. If allowed to rest for even a few days the ration must be reduced proportionately. Exercise must be given to keep the circulation of the limbs active, and no more food given than suffices for actual daily requirements.

d. Thin animals.—Thin animals should be fed grain four or more times daily and variations made in the ration to tempt the appetite. Steamed oats as an evening feed daily or every second day are useful for such cases, as is also a small bran mash once or twice weekly. As the animal gains in strength, the quantity of each feed may be increased. Water should be supplied in unlimited quantities.

111. Care of forage.—Dampness decays forage in storerooms. Hay and grain rooms must be kept dry and ventilated at all times. Forage should not be piled on floors nor in contact with walls. Boards properly spaced so that the air may circulate under the pile should be used as a foundation for each pile. Spaces should be left between piles and between each pile and the walls and ceiling in order that air may circulate freely around them. Grain should be protected from mice and rats.

112. Feeding rules.—*a.* Water before feeding or not until at least one hour afterward.

b. Never feed grain to heated animals; hay will not harm them.

c. If expecting hard work immediately after feeding, give only a half feed.

d. Do not work the animal immediately after a full feed.

e. Use feed bags when feed boxes are not available.

f. Give a small feed of hay before each grain feed.

g. Feed a bran mash occasionally as the general condition of the animal or the condition of the droppings may indicate; or feed daily from 1 to 2 pounds of dry bran mixed with the oats.

h. Feed salt.

i. Never feed hay on the ground if it can be avoided.

j. Remove feed bag as each animal finishes feeding.

k. Keep feeding utensils clean.

l. Feed clean food.

m. Watch your horses feeding and know which are the slow and the shy feeders.

n. Watch their condition and fix the ration for each horse according to his needs.

o. Graze whenever the opportunity occurs.

p. Encourage men to bring handfuls of grass or other food to their horses.

q. Be economical in feeding; waste nothing; let the horse get the full value of his ration.

r. Feed most of the ration, especially the hay, at night. The animal then has plenty of time to eat and digest food before working again.

s. Feed at regular hours each day.

t. Feed grain at least three times daily and to thin horses four times.

u. Feed grain in small amounts and often.

113. Feeding schedule.—The times of feeding and watering should be definite and regular, and organization commanders should prepare a schedule for this purpose. The amounts of grain to be fed at each feeding should be marked on a suitable card or blackboard on the heel post in the rear of each horse.

114. Watering.—*a. General.*—(1) Under average conditions animals require about 8 gallons of water per day. Excessive heat and heavy work will double ordinary requirements.

(2) Animals suffering from communicable diseases should always have a separate watering place. When practicable, such animals should be watered from individual buckets.

(3) During watering there must be no shouting or confusion of any kind.

(4) Restless and kicking animals should be watered separately.

(5) Never allow an exhausted animal to drink his fill at once, especially if he has gone long without water. Give him a little at a time at 5- or 10-minute intervals until he is satisfied.

b. In camp or garrison.—(1) In camp or garrison, animals should be watered at least three times daily. At work, especially when the weather is warm, four times daily should be the rule. Water should be given previous to feeding.

(2) Animals should not be watered for at least an hour after feeding, but they may be allowed to drink freely while at work, even though sweating. If brought in hot, they may be given a few swallows at intervals, but should be kept moving until they have cooled down.

(3) Water only such number of animals as the trough will accommodate. They should be allowed ample time to drink their fill and not be led away the first time they raise their heads from the water. Animals should be led to and from water at a walk and, if possible, not more than two animals per man.

(4) If a stream is the source, care must be exercised not to foul the water for succeeding units. (See *c* (3) below.)

c. On the march.—(1) Animals should be watered within reason whenever opportunity occurs, especially on hot days. The usual method of watering before feeding will be adhered to, but if a stream is crossed an hour after feeding they may be allowed to drink if circumstances permit the delay. Bits, especially the curb, should be removed when it is intended to give a full watering. To prevent chilling or digestive disturbances, animals watered on the march, especially if sweating, should continue to march at once at the walk for a period of 10 or 15 minutes before halting or taking up the increased gait.

(2) When water is scarce its issue must be carefully regulated to avoid waste. Animals can drink from a very shallow vessel if their bits are removed. A very small quantity will revive overtaxed animals; this should be given in small rations rather than one long draft. When the amount is very scanty it is best to offer the exact quantity to each animal in turn.

(3) While watering, overcrowding of the animals should be prevented (normally about 2½ yards of water space per animal) and plenty of time allowed for each animal to drink his fill. Groups of animals should come up together and leave

together. One man should not lead more than two horses to water and should stand between them while they are drinking. Do not tie their halters together. After they have finished drinking, animals are inclined to paw the water and even lie down. This must be prevented, else the water will be rendered unfit for commands which follow.

(4) If practicable, the place for the noon halt should be 2 or 3 miles beyond a convenient watering place. The animals should be watered and then moved forward to the designated resting place. Feed a little hay to the animals as soon as they are tied on the picket line. Animals should not be watered until they have been at least 45 minutes in camp.

Section III

STABLE MANAGEMENT

115. General.—*a.* The organization commander is responsible for the proper performance of stable duty in his organization. He should see that the men on duty at the stables are properly instructed and that the stable routine prescribed by him is faithfully carried out. An officer should habitually be present during stables.

b. The animals are assigned to stalls and to places on the picket line by platoons or sections. During the day, except in very cold, very hot, or stormy weather, they should stand at the picket line or be at liberty in the corral.

116. Floors.—*a. Cleanliness.*—The floors and aisles of stables should be thoroughly cleaned daily. During warm weather, brick, cement, and stone floors are best cleaned by scrubbing with water, using the hose and stable brooms. Earthen floors at all times and hard-surfaced floors in cold weather are cleaned by scraping with a shovel or hoe and sweeping with a stable broom.

b. Slippery floors.—If the floors wear smooth and become slippery, the surface should be roughened by chipping with a cold chisel and a mallet. If slippery from ice, the surface should be sprinkled with sand or sifted ashes before the bedding is put down. Likewise, when necessary, the aisles are covered temporarily with bedding and the approaches with sand and cinders to prevent the animals from slipping.

c. Maintenance of earthen floors.—Earthen floors require continual work to keep them clean and smooth. As soon as wet depressions occur they should be dug out and filled with fresh clay bound with cinders or crushed rock. The floor at the front of the stall for a distance of about 2 feet from the rear edge of the manger should be level in order that the animal's front feet may rest on a level surface. Back of this space the floor should slope gently to the rear to provide drainage. A drop of about 2 inches is sufficient.

117. Mangers.—*a. Cleanliness.*—Feed boxes and mangers should be brushed out daily, care being taken to see that all particles of food are removed from them and from the woodwork surrounding them. Advantage should be taken of exposure to sunlight and fresh air to keep feed boxes clean. In addition to the daily police, a monthly cleaning of feed boxes and the woodwork immediately surrounding them should be made by scraping and scrubbing with soda ash and hot water. Feed pans or boxes should be washed immediately after each bran mash feeding.

b. Cleaning solution.—A solution of 5 pounds of soda ash to 12 gallons of water is used. To clean feed boxes this mixture should be placed in a receptacle large enough for the purpose and brought to the boiling point. After being cleansed and all food particles removed, the feed boxes should be placed in the boiling solution for a period of from 3 to 5 minutes and then removed and thoroughly rinsed. This is best done by the use of a hose. None of the soda-ash solution should be allowed to remain upon a feed box. Care should be exercised by those handling the solution to prevent its coming into contact with the skin or mucous membranes, particularly of the eye.

118. Woodwork.—*a.* There should be no sharp or rough projections in the woodwork of the stables on which animals might injure themselves. All fittings should be flush with the surface to which they are attached. When used, kicking bars should be adjusted to such height that they are level with a point midway between the hock and stifle. The rear end of the bar should be equipped with a release which will operate when an animal gets a leg over the bar. The top bar of the manger should be wide enough that an animal cannot readily grasp it with his teeth. The top bar of the manger and other exposed edges of wood should be covered with metal of moderate thickness to prevent gnawing of woodwork or "cribbing." Woodwork about the manger can be cleansed best if the wood is left unpainted. Woodwork about manger and front of the stall should be cleansed by scrubbing as frequently as may be necessary to maintain the wood free of stains or incrustations of dirt. Woodwork of that part of the stall which the animal cannot reach with his mouth may be left natural or painted and should be washed frequently to remove dust and stain. Whitewash is not suitable for use in stables.

b. Three tablespoonfuls of lye in a bucket of warm water makes a good solution for cleaning the coating of dirt from the mangers and woodwork of the stalls and improves the appearance of the stables. Care should be exercised when using this solution that none of it gets on forage or bedding.

119. Water troughs.—Water troughs should be emptied and thoroughly cleaned each morning. A water trough placed in a corral should be kept plentifully supplied with water at all times when the animals are in the corral. Strange animals should not be permitted to drink at the water troughs or be fed in the stables. One animal with a contagious disease may affect all the animals of the organization. Animals suffering from contagious disease should be watered from buckets.

120. Corral.—*a.* The corral should be cleaned at least once a day. Thorough drainage should be provided and all depressions kept filled so that pools of water will not form after rains.

b. The corral fences should be of sufficient height and stoutness to restrain animals and should be kept in a thorough state of repair at all times. They should be entirely free of all rough or sharp projections.

c. The picket line and corral should be sprinkled to keep down the dust. Crude oil or used oil drained from crankcases of motors may be used to advantage. If there is no shade, shelter from the sun should be improvised, as animals lose condition rapidly if forced to stand in the hot sun for long periods of time.

121. Picket line.—*a.* In garrison or permanent camp the picket line should be elevated about 5 feet above the ground on heavy posts. The picket line may be made of 1½-inch rope or small wire cable. It must be kept taut at all times. The length of the line should provide 4 feet of standing space for each animal. If necessary to locate the line on a considerable slope, the line should run up and down the slope instead of on a contour around it.

b. A second line below the first, breast high, is excellent for keeping the animals all on one side. It prevents fighting across the line and the tangling of animals.

c. The standings along the picket line should receive the same attention and care as stable floors. A foundation of stone should be provided with a covering of finely crushed rock, gravel, or cinders, well-bound and rolled firm. In wet climates the standings should be raised above the ground level.

122. To secure animals.—*a.* The proper length of rope for tying an animal in a stall is such as will permit him to reach all parts of his manger and to lie down with ease. The length of the permanently installed tie rope should usually be such that the snap hangs 2 or 3 inches above the floor. Permanently installed tie ropes of proper length will prevent many injuries due to soldiers tying too long or too short or so insecurely that animals get loose. Ropes and not chains should be used for this purpose. The rope is best fastened in the tie ring by splicing.

b. If animals are to be fed hay while on a picket line, the length of rope should be no more than enough for them to reach the ground comfortably. If animals are not to be fed on the line, a much shorter length of rope should be used.

c. Any knot which is perfectly secure and capable of being untied rapidly may be used for securing animals.

123. Fire prevention.—*a.* All stables must be provided with fire equipment and knives for cutting tie ropes. An extra tie rope for each horse, for use only in case of fire, should be conveniently available. Periodic fire drills should be held. Smoking in stables must be strictly prohibited.

b. Animals will not leave a burning stable of their own volition. Each horse must be led to safety and secured in some nearby stable or corral or tied on a picket line at a safe distance.

124. Ventilation.—Animals rarely suffer from cold air, provided there is no draft. They do, however, suffer greatly from foul air. Ventilators and louver boards in the stable roof carry off the warm, tainted air. Fresh air is obtained through windows and doors. Windows should hinge at the bottom and open inward, thus forcing the incoming fresh air up and over the animals instead of directly on them. Doors should be closed on the windward side in stormy weather.

125. Bedding.—*a.* A good bed induces animals to lie down, and it also prevents injuries to elbows and other parts in contact with the stable floor. The bed should be level, dry, warm, and elastic and have a clean surface for the animal to lie upon. Clean straw is the best material for bedding.

b. In the morning the bedding should be removed from stalls and well shaken out, and only the dung and short or badly soaked refuse taken to the stable dump. That part of the bedding that can be used again should be spread out in the aisle to dry. The floors of stalls should be thoroughly cleaned each morning and allowed to dry out during the day.

c. As late as possible in the afternoon, the dried bedding is mixed with such new bedding as may be necessary and the stalls bedded. The bedding should be evenly spread over the bedded portion of the stall. As horses usually lie well back, the bedding should not be laid right up to the front of the stall.

126. Grooming.—Grooming is essential to the general health, condition, and appearance of animals. The value of grooming depends upon the thoroughness and speed with which it is done. Men should be encouraged to work hard and rapidly and to do a thorough job in a minimum time. Each man's grooming should be inspected upon completion and, if satisfactory, the man dismissed. Grooming by squads is an efficient method and develops leadership in squad leaders. Efficient grooming is best obtained when the individual soldier takes a personal pride in the appearance of his mount or team.

a. Time to groom.—(1) Every animal should be groomed thoroughly at least once each day. Before leaving the stable for work or exercise, horses are brushed off; mane, if any, neatly arranged; tail brushed out; eyes cleaned; nostrils and dock wiped out; and feet cleaned.

(2) On return from work or exercise, grooming should be done immediately after equipment is wiped off and put away. Heated, wet, or sweating animals should be cooled out before being groomed. In the case of such horses the equipment should be removed and quickly put aside, then the horse given a brisk rubbing with drying cloth to dry the coat partially. He should then be blanketed and walked till cool.

b. Equipment.—(1) *Contents of kit.*—Each mounted man should have a grooming kit equipped with the following:

(*a*) *Currycomb.*—The currycomb is used to remove caked mud, to loosen matted scurf and dirt in the hair, and to clean the horse brush. It produces the best effect when applied gently in small circles. It should never be used on the legs from the knees or hocks down nor about the head.

(*b*) *Hoof hook.*—The hoof hook is used to clean out the feet.

(*c*) *Horse brush.*—The horse brush is the principal tool used for grooming. When properly used it reaches the skin, the bristles or fibers of the brush penetrating through the hair of the coat.

(*d*) *Grooming cloth.*—The grooming cloth is used to clean out the body orifices and to polish the coat. It is made from old toweling or condemned blankets, about 2 feet square.

(2) *To wash and disinfect tools.*—Grooming tools should be washed and disinfected occasionally as a precaution against the spread of skin diseases. Clean in soda-ash solution (mess-kit cupful to two-thirds of a galvanized iron bucketful of water). Immerse in a 3 percent cresolis solution (15 mess-kit spoonfuls of cresolis to two-thirds of a galvanized iron bucketful of water) for 30 minutes. To stiffen the bristles of brushes after disinfection, soak in strong salt solution (add salt to water until no more salt can be dissolved) for a few minutes and stand with bristles down to dry. Grooming and drying cloths should be washed with soap and water once a week.

c. Process of grooming.—(1) Clean out the feet thoroughly, working from the heel toward the toe. A thorough cleaning out is necessary to the bottom of the commissures or depressions between the frog and the bars. The deepest part of these depressions is near the heels and is the usual seat of thrush. Care should be exercised in cleaning the clef. It should not be deepened by cutting the horn of the frog. Inspect the feet for thrush, torn frog, loose shoes, etc., reporting defects at once to the stable sergeant.

(2) Take the currycomb in the right hand, fingers over the back of the comb, and the brush in the left hand; first use the currycomb on the left side of the animal, beginning on the neck, then breast, withers, shoulders, foreleg down to the knee; then back, side, belly, croup, and hind leg to the hock. Strike the currycomb against the heel frequently to free it from dirt.

(3) Brush entire left side of the animal in the same order as when the currycomb was used except that in brushing the legs brush down to the hoof. After every few strokes clean dust and hair from the brush with the currycomb. In using the brush the man should stand well away from the animal, keep his arm stiff, and throw the weight of the body against the brush. In grooming the belly apply the brush the way of the hair.

(4) Pass to the right side of the animal, change the brush to the right hand and the currycomb to the left hand, and groom in the same order as above.

(5) Brush head, mane, and tail. In cleaning mane and tail begin brushing at the ends of the hair and gradually work up to the roots, separating the locks with the fingers so as to get out scurf and dirt. Tails require occasional washing with warm water and soap.

(6) With grooming cloth wipe the eyes, nostrils, and lips; rub the head, ears, and muzzle; clean the dock and outside of the sheath, and give a final polish to the coat.

127. Hand rubbing.—*a.* Hand rubbing is restful to tired muscles, stimulates the circulation, removes loose hair, and helps to produce a glossy coat. The hands are slapped down briskly on the coat, one after the other, with the weight of the body behind them, then moved over the skin with firm pressure.

b. Upon unsaddling after a march, rub the back with a circular motion to restore circulation.

c. After hard work or at the end of a march the legs below the knees and hocks should be hand rubbed and carefully examined for evidence of injury.

128. To clean the sheath.—The sheath requires occasional cleaning with warm water, a sponge, and a mild soap to remove accumulated secretions and dirt. Form a lather with the sponge and carefully remove the accumulated secretions. The ball of waxy secretion called a "bean" which develops in a depression in the head of the penis should be removed with the fingers.

129. To pull manes and tails.—*a.* Manes and tails should be pulled thin and even by grasping a few hairs at a time and sliding the hand up close to the roots, then giving a quick jerk so as to pull the hairs out by the roots. Always work on the longest hairs and on the under side of the mane and tail.

b. Tails should be shortened to about 4 inches below the hocks.

130. To trim manes and fetlocks.—*a.* Fetlocks should be neatly trimmed, and that part of the mane on which the crown piece rests should be closely clipped. The hair on the back surface of the pasterns should be clipped or trimmed shorter than one-quarter inch.

b. Clipping manes is strongly recommended, especially for field service. When manes are clipped, the clipping should be done close all the way up.

131. Clipping.—*a.* As a rule the clipping of working animals is recommended. Whether clipping is advisable depends upon the nature of the coat, the climatic conditions, the amount and character of the work to be performed by the animal, the character of the stables, the amount of clothing, the availability of personnel, time for grooming, etc. Clipping is not recommended under field conditions during moderately cool or cold weather. During the severe weather in colder climates it is not advisable to clip the legs. Where animals are to receive considerable work under the saddle it is advisable to leave a saddle patch the size of the blanket under the saddle.

b. If clipping is practiced, it should begin in the fall before the winter coat becomes heavy. Animals should be reclipped during the winter as often as the length of the coat warrants it. Clipping under most conditions should cease as soon as spring shedding begins. Clipped animals should be warmly clothed and not exposed to low temperatures in corrals or stables. When animals are reclipped during cold weather it is advisable to exercise the animals immediately after clipping until a light sweat appears and then dry, groom thoroughly, hand rub the body, and blanket. Clipping lessens the labor of grooming but the clipped animal needs the same thorough and vigorous grooming as an animal in full coat.

c. Power clippers are best for clipping animals, though hand clippers will serve the purpose. Clipper heads and blades must be used with care, as they are easily dulled and broken. They can be resharpened and should be as often as necessary. Grooming should precede clipping, as many blades are broken by sand and dirt in a dirty animal's coat. While using power clippers a small pan of kerosene oil should be at hand and the clipper head immersed frequently with the machine running.

132. To wash animals.—Washing of animals is strongly condemned. When necessary to wash the lower leg or other parts of the body to remove stains, thorough drying of the wetted hair and skin is necessary to prevent scratches or chilling.

133. Blankets.—*a*. Blankets are used for the following purposes:

(1) As a protection against cold and storm.

(2) To cool out wet or heated animals.

(3) To improve the coat.

(4) To protect thin-skinned and clipped animals from flies.

(5) As a protection against chill in the case of sick animals.

b. Both lined and unlined blankets are issued. A serviceable fly sheet can be made from grain sacks. Horse blankets should be worn while at rest only.

c. Blankets must be properly adjusted and secured. Otherwise they are likely to slip off and become torn. All straps and fastenings must be kept in good repair. Blankets must be kept clean. Those used on animals suffering from communicable diseases must be thoroughly disinfected before being used on other animals. To assist in the prevention of possible spread of skin diseases or parasites, it is advisable to stencil on the blanket the name or Preston brand number of the horse and confine its use to that animal only.

134. Exercise and conditioning.—*a*. Animals are considered to be in fit condition when the body and muscular organs perform without injury the work required. Fit condition is accomplished by systematic exercise and feeding, which must be continued even after the condition has been attained. Endurance, stamina, a good state of flesh, and resistance to disease are necessary. These qualities can be attained only by intelligent feeding and long periods of conditioning work at the slower gaits, largely the walk. The riding animal must be conditioned to carry the weight of full field equipment.

b. The exercise must be regular and graduated and always within the limits of the animal's strength. Working tired animals when unfit is very likely to cause accidents or disease.

c. If animals are not worked for any considerable time the process of conditioning must be repeated. The minimum amount of systematic exercise necessary to maintain animals in working condition is 2 hours daily. It is not desirable to maintain animals at all times at the peak of condition for field duty, but rather to have them in such condition that a relatively short period of carefully scheduled work will put them in fit condition for the anticipated duty.

d. The kind and amount of exercise given to animals depend upon the work they are to perform, the condition of the animals, and the number of men available to give the exercise. When there is one man available to exercise one or two animals, the most satisfactory exercise may be had by organization under the supervision of the troop officers. But when there are many animals and few men, special means must be adopted.

e. Exercise tracks or rings are probably the most economical means of exercise, considering man power, but there is more danger of accidents than when riding one animal and leading another. A circular or oval track about 15 feet wide and inclosed between fences 5 or more feet high is all that is necessary. Animals should be admitted to the track in groups of not more than 24 at a time. There must be a rider to lead and one to follow each group. By this means the gait can be regulated and stampeding or crowding eliminated. Turning animals loose in corrals and driving them around is not recommended, as too many accidents may occur.

f. Any form of exercise should begin with a period of 10 or 15 minutes at a walk so that good circulation may be established. It should end at a walk in order that animals may be returned to the stables dry and breathing normally. Dirt roads should be used habitually for exercise, with only enough work on hard roads to accustom the animals to traveling on them. When one man is exercising two horses he should ordinarily ride one out and the other in, and he should lead each horse alternately on his right and left. Playful or vicious animals should be led with a very short rein to prevent accidents.

g. The walk is the prime conditioning gait. Walking develops muscle, while trotting and galloping develop wind, but if either of the faster gaits is used to excess the animal will lose rather than gain condition. The troop horse gets enough fast galloping during the training of men and maneuvering, and his exercise should be carried out at the walk and trot. The length of trot periods must be suited to the condition of the animal but in no case should they exceed 7 minutes.

h. The routes selected for exercise should be varied from day to day to avoid monotony. It is best to return by a different route from the one taken going out so that animals do not fret when they are headed toward stables.

i. Draft and pack animals should do a considerable part of their exercise drawing loads and carrying packs so that their muscles may be hardened to the work.

135. Care of the feet.—Every officer in charge of animals must continually watch the condition of the animals' feet and see that they receive proper care. Every man who rides or drives an animal must do the same, reporting promptly to his squad leader or the stable sergeant any need for attention. (See also sec. VI.)

136. Stable vices and their prevention.—Stable vices are objectionable habits of animals, practiced while idle in the stable or on the picket line, and are caused usually by insufficient work or exercise. As a general prevention to their contraction, animals should be kept out of the stables as much as possible during the day, either at work, on the picket line, or loose in the corral.

a. *Weaving*.—This is a nervous habit in which the animal rocks on his forequarters, constantly shifting his weight from one forefoot to the other, and swaying the head and neck from side to side. Many horses stop this habit when receiving sufficient work and when stabled so that they are in close company with their stable mates.

b. Cribbing and windsucking.—These two habits are closely related; in fact, they are but variations of the same condition. Both habits are caused largely by idleness and are incurable, but the practicing of the habit in many cases may be prevented. The cribber catches some stationary or resisting object with his incisor teeth and pulls or bites while arching the neck and drawing the muzzle in toward the breast. As animals become more confirmed in this habit, while practicing the act, they will give a sort of gulp associated with a distinct grunt and pass air into the stomach. Such animals are called "windsuckers." Cribbing or windsucking may be prevented by stabling in a stall without projecting fittings. Another method is to pass a narrow strap just in front of the ears and around the throat just back of the jaw and buckle it in this position quite tightly so that when the head is flexed downward the animal is uncomfortable. Chronic indigestion, general debility, and colic frequently result from these vices.

c. Biting.—Biting is usually the result of animals being teased by men in a playful way, but once acquired is a dangerous vice. As a prevention, teasing must be stopped and, if the animal is dangerous, only one man should handle him. The danger may be lessened by use of a muzzle, a thick wooden bit, or a side stick. Swift, painful punishment will often stop the habit.

d. Kicking against stall.—This habit is usually acquired through lack of work. It may be cured by using a short chain, one end of which is fastened to a hobble around one rear pastern, the other end attached to a small wooden ball. Padding the stall will prevent the animal from injuring himself and will very often stop the habit, for many animals kick the stall for no other reason than to hear the sound produced.

e. Gnawing woodwork.—This is a restless habit acquired by animals kept tied up without work or when deprived of hay. The cure is work and plenty of hay. The prevention is to wrap the woodwork with wire or cover it with heavy sheet metal.

f. Eating dung.—This habit is a morbid appetite resulting from poor condition, caused by chronic indigestion. Animals in fit condition seldom form the habit. The only prevention is to keep the animal in good condition, well-fed and exercised, and tied so that he cannot reach his own or other animal's droppings.

g. Greedy feeding.—A greedy feeder eats as fast as he can, grasping huge mouthfuls at a time and throwing his feed with his muzzle out of the box on the ground, and later eating the dropped grain. This results in indigestion, the bolting of feed, eating dirt and trash, and is wasteful on account of the amount of forage lost. To prevent this vice, place several large stones in the feed box, feed the grain with chop, place a wire screen with half-inch mesh inside of the box and resting on the grain, or feed in a wide-bottomed feed box. If solid bottomed mangers are used, the best solution is to feed the grain on the manger bottom instead of out of separate feed boxes.

h. Tearing blankets.—The only prevention of this habit is the use of a leather muzzle or side stick.

i. Halter pulling.—This vice is usually contracted by the use of weak halters or ropes which permit the animal to break away when frightened. Tying with the reins has the same effect. As a prevention use a very heavy neck strap and tie rope. The vice is sometimes cured by tying a piece of rope around the animal's body just in rear of the forelegs, from which the tie rope is run between the legs through the halter ring and to the manger or picket line. After a few attempts to pull away, the animal will probably give up. This should be applied for several days and then just a light cord tied around the body as a reminder. Another method often successful is to pass the free end of the tie rope from the halter through the manger tie ring and secure to a hobble placed on one fore pastern.

137. Restraint and control of animals.—In the management of animals, restraint is sometimes necessary. Always select the mildest and least dangerous method that will accomplish the end sought. Kindness, perseverance, and tact will often accomplish the desired purpose without resort to special means of restraint.

a. Twitch.—(1) The twitch is the simplest, handiest, and most common method of restraint. As it shuts off circulation in the lip, it should never be used continuously for an extended period of time and never with greater force than is actually necessary.

(2) A twitch is made by running a small piece of rope or rawhide through a hole in the end of a rounded piece of wood 2 to 5 feet long, such as an ax handle, and tying it into a short loop.

(3) The loop of the twitch is passed over the upper lip, which is seized by the hand and drawn forward, care being taken to turn the edges of the lips in so as to prevent injury to the mucous membranes. The cord is then twisted by turning the stick until sufficient pressure is obtained.

b. Side stick.—This method of restraint is used for vicious animals to prevent them from biting men while grooming or to prevent an animal from doing injury to a wound. The side stick is made from a stout wooden rod from 3½ to 4 feet in length, with a small cord or strap at each end, one for attaching to the noseband of the halter, the other to the surcingle.

c. Muzzle.—This means of restraint is useful to prevent an animal from eating bedding or chewing or tearing the dressing from a wound. Muzzles are best made of leather. They should be provided with attachments for fastening to the halter rings or with straps which are sewed to the muzzle and fastened over the poll.

d. Cross tie.—The cross tie consists of tying the head in an elevated position with two tie ropes. The ropes extend from the tie ring in the halter to opposite sides of the stall. The cross tie is useful to prevent an animal from chewing a wound or dressing and from lying down when it is desired to keep him standing.

e. Blinds.—Blinds are used to restrain nervous or vicious animals by depriving them of sight. A blind may be made in the form of a hood or of a piece of leather broad enough to cover the eyes and provided with strings for fastening to the halter.

f. Knee straps.—In shoeing an animal or when operating on him it may be necessary to fix a foreleg. A strap or rope is used to fasten the pastern of the foreleg. The leg is first well bent at the knee. A rope or strap is then attached to the pastern by means of a loop and the free end passed over the forearm and back to the pastern, drawn tight, and tied or buckled.

g. Side line.—For securing a hind leg the side line is used. It is made of three-quarter-inch rope about 20 feet long and a leather hobble provided with a ring. The hobble is placed around the pastern of the leg to be raised and buckled snug. One end of the rope is secured around the animal's neck by a nonslip tie and the free end passed through the hobble ring and back through the loop. The leg is then brought forward by pulling on the free end of the rope and held in the desired position by an attendant.

h. Casting rope.—(1) The casting rope is made of three quarter-inch rope about 50 feet long. This is doubled and from 3 to 4 feet of the doubled end tied in a nonslip loop which is passed over the animal's head and adjusted about the neck. The free ends are passed backward between the forelegs and around the rear pasterns, or through hobbles if used, and thence upward and back through the loop at the neck. A strong man should hold the animal by the head, standing on the side on which it is desired to throw the animal. When standing on the left side, the left hand should grasp the halter firmly, the halter shank, brought over the top of the neck, being double, held in the right hand. Two or more assistants grasp the free end of the casting rope and by a quick pull to the rear take the animal off his feet. Acting in conjunction with them, the man at the head pulls the head and neck to the ground.

(2) When the animal is down, the man at the head should hold the head firmly against the ground by placing one knee on the neck and raising the animal's muzzle from the ground. At the same time the attendant handling the ropes on the side not against the ground should quickly draw the legs up against the side and secure them by a double half-hitch made of the free ends of the rope. When one side is thus secured the animal may be rolled over and the other legs tied in a similar manner.

(3) Casting can be done with least effort if one foreleg is secured by means of a knee strap (*f* above). When necessary to pass ropes around the rear pasterns (in case hobbles are not available), always bandage the pasterns to prevent rope burning.

91

138. Assistance to cast animals.—*a.* In going to the assistance of a horse or mule which is down and unable to rise without aid, the following procedure is suggested:

(1) See that the animal has plenty of room. If it is in a single stall, pull it out into the aisle.

(2) Turn it over by attaching ropes to the lower hind feet and forefeet. Then see that the feet are on the same level with or lower than the back.

(3) Assist the animal to roll up on its breast and allow it to remain quiet for a few minutes.

(4) Pull the front feet out in the position naturally assumed by a horse when it is about to rise.

(5) Stand close against the buttocks and grasp the tail. Be prepared to exert a pull upward and forward. Leave the head alone.

(6) Now speak sharply to the animal or slap it lightly with a strap and lift the tail as directed.

b. In case it is impossible to raise the animal by the above means, use a sling. The articles required are a block and tackle of sufficient strength and 30 feet or more of rope. The block may be fastened above the animal in the stable, or in the field a tripod may be improvised from strong poles. The rope is placed around the animal as follows:

(1) Double it, put the head through the loop, and carry the loop back to the collar seat. Pass the free ends of the rope between the front legs, crossing the ropes just before they go between the legs. Pass the lower rope under the body so that it emerges just back of the withers.

(2) Lay the upper rope over the chest so that it crosses the lower rope at this point.

(3) Bring the lower rope over the body and between the hind legs.

(4) Pass the upper rope under the body in a position corresponding to the other rope, crossing the two ropes as they emerge from between the hind legs.

(5) Bring both ropes up over the buttocks, one on each side of the tail, and pass them forward along the back, under the crossed ropes, and tie to the loop which lies at the top of the neck. Draw the ropes as tight as possible before the knot is tied. Now attach the hook of the block under all the ropes at the point on the back where they intersect, the single

strands crossing to go between the hind legs and the double strand coming forward from the tail, then raise the animal.

139. Records.—*a.* The following records are kept at each stable by the stable sergeant:

(1) List of public animals by Preston brand number or a file of copies of the Horse Record Cards (W. D., Q. M. C. Form No. 125). The original W. D., Q. M. C. Form No. 125 for each animal is kept in the unit supply office.

(2) Record of stable property.

(3) Forage record.

(4) Shoeing record.

(5) Morning report of animals.

(6) Sick report of animals.

b. Descriptive cards of public animals are made out, one for each animal, on W. D., Q. M. C. Form No. 125. The form gives the name and number of the animal, age, sex, color, and markings. It should contain a complete record of the service of the animal from the time of purchase until finally disposed of.

c. All stable property and equipment are listed in a property record book, showing each class of tool or equipment separately.

d. An accurate record of all forage received, on hand, and consumed should be kept posted accurately from day to day. The form shown below may be used for this purpose:

Forage record

(Organization)

------------------------------ --------
(Month) (Year)

Date	Number of animals	Grain on hand	Grain received	Grain consumed	Grain balance	Hay on hand	Hay received	Hay consumed	Hay balance	Remarks
--------	------	------	------	------	------	------	------	------	------	----------
--------	------	------	------	------	------	------	------	------	------	----------
--------	------	------	------	------	------	------	------	------	------	----------
--------	------	------	------	------	------	------	------	------	------	----------

e. An accurate record of the shoeing of all animals of the organization is kept as shown in the form following.

Shoeing record

(Organization)

(Year)

Animal's No.	Animal's name	January	February	March	April	May	June	July	August	September	October	November	December	Remarks
T-10	Murat	5	9	11	10	15	- - -	- - -	25	20	18	20	16	Hind shoes removed and not replaced March 11. All shoes removed May 15, sick in stables May 15 to August 25.
		10	- - -	- - -	30	- - -	- - -	- - -	- - -	28	- - -	- - -	- - -	

Explanation: The circle appearing in the rectangle for any month represents the body of the horse. The lines outward from the circumference represent the legs of the animal. The number in the circle represents the day of the month. Thus, in the case of T-10, "Murat," this horse lost the shoe from his right fore on January 5 and the record shows that it was replaced that date. On January 10 his other three feet were shod. On February 9 he was shod all around. On March 11 he was shod in front and the remark shows that his hind shoes were pulled and left off. On April 10 he was shod in front. On April 30 he was shod behind. On May 15 all shoes were removed and not replaced. On August 25 he was again shod all around, etc.

f. A record is kept of all animals in the stables, showing the changes that occur from day to day. The form shown below, painted upon the bulletin board at the stable, is used for this purpose:

Morning report of animals

(Organization)

	Duty	Hospital	Sick in stable	Being shod
Public horses----------	------	----------	--------	--------
Private horses----------	------	----------	--------	--------
Mules----------	------	----------	--------	--------

Changes

Stable sergeant

g. An animal sick report is kept at each stable. The following form in a book is recommended:

Animal sick report

(Organization)

From---------------------------------
(Date)

To---------------------------------
(Date)

Date	Brand	Name	Diagnosis	Status	Remarks
----------	------	----------------	--------	------	------------------
----------	------	----------------	--------	------	------------------
----------	------	----------------	--------	------	------------------
----------	------	----------------	--------	------	------------------

Section IV

FIELD MANAGEMENT

140. Care of animals on the march.—*a. Rate and length of march.*—(1) When properly conditioned, cavalry can march on average roads at a rate of 6 to 6½ miles per hour up to about 30 to 35 miles as a day's march. Cavalry can maintain this rate and daily length of march, with one day's rest per week, for an indefinite period of time. Horse artillery can travel at the same rate as cavalry. For ordinary marches, horse-drawn artillery marches at the rate of 4 to 5 miles per hour on good roads. Under average conditions, wagon transportation marches at a rate of 3 to 4 miles per hour. As a rule the rate of march varies inversely with the size of the command. The above approximate daily lengths of march may be considerably increased for marches of but a few days' duration.

(2) To fit animals properly for long marches, a period of planned conditioning is necessary. If mounted commands are called upon to perform active campaign duty without a warning period in which animals may be conditioned, the early work should be graduated to the fitness of the animals with the aim of conditioning or "seasoning" them during the early stages of campaign.

b. The start.—When practicable, daylight marches begin as early as is consistent, allowing ample time for the men to breakfast, animals to feed, and transportation to be loaded. In the combat zone, night marches will be the general rule and all commands should receive training therein.

c. Gaits.—The first 10 minutes after leaving camp should be at a walk. The trot should be used only on level ground or down gentle grades. Trot periods should not exceed 7 minutes at one time. Frequent changes of gait are desirable. The canter or gallop should not be used as a marching gait except when the tactical situation demands it. It is advisable to lead for short periods (3 to 4 minutes) immediately before and after halts. Leading should be resorted to when steep up or down grades are encountered or whenever the condition of animals makes it advisable.

d. Formation.—(1) Marching mounted troops on both sides of the road promotes the comfort of the horse and rider. Riding animals should keep off hard-surface roads if suitable footing is available at the sides of the road. Vehicles should always keep to the firmest and most even part of the road.

(2) If the military situation permits, commands should march with sufficient distance between subordinate units to allow the dust to settle or blow away and to take advantage of terrain conditions for appropriate changes of gait.

e. Halts.—(1) A halt of 10 or 15 minutes should be made at the end of the first 45 minutes of marching to permit troops to relieve themselves, adjust equipment, and inspect shoeing, loading, etc. Subsequently a halt of 5 or 10 minutes should be made hourly.

(2) It is desirable to finish the day's march as soon as practicable; long halts in the course of a daily march are not made unless special conditions require it. This applies to marches of less than 5 or 6 hours' duration. For marches of over 5 or 6 hours' duration, or under conditions of excessive heat, a halt of 1 to 4 hours should be made about the middle of the day.

(3) At halts, animals should be permitted to graze. During long halts, saddles, packs, and harness should be removed. Pack animals should be relieved of their pack loads at every opportunity. At each hourly halt, saddles and packs should be adjusted, men working in pairs. Feet should be examined at each halt for evidence of loose shoes or rocks lodged in the feet. Officers and noncommissioned officers should inspect the animals of their unit at each halt.

f. Watering.—See paragraph 114*c*.

g. Feeding.—The principles of feeding animals on the march are the same as for feeding in camp. (See pars. 108 and 109.)

141. To cross streams.—*a. Fords.*—(1) When crossing a stream at an unfamiliar ford or where there is no ford, the crossing should invariably be examined before committing the command to it. The sound sand footing of some safe fords begins to wash away after several animals in column have passed, and the ford rapidly deepens. In such cases the crossing should be made on a wide front.

(2) Deep streams should be crossed on a relatively wide front to prevent damming; streams with treacherous bottoms, on a narrow front with limits of the ford marked.

(3) Animals should be allowed to drink before getting well into a dangerous ford, but having been committed to the crossing must be pushed vigorously ahead and not allowed to stop. Riders should take their feet out of the stirrups if the current is swift. Eyes must be fixed on the point of the opposite bank toward which riding and not on the water.

(4) Quicksand is dangerous and frequently shifts. Crossings on this type of bottom should be avoided.

b. Ice.—(1) Ice, if thin or rotten, is a serious obstacle to crossing a stream. If thick and sound, it is a very good bridge in itself. When time permits, the crossing should be sanded. Boats used in ice must be protected with chafing pieces, especially near the water line at the bows. Heavy ice, floating rapidly, makes a crossing impracticable.

(2) On sound ice, mounted troops may cross with large intervals on a thickness of 4 inches, and light artillery on 6 inches.

(3) Loads may be carried on ice of less thickness or on unsound ice by distributing the weights. Wagon beds may be placed on boards and used as sleds to cross supplies. Animals may be hauled across on platforms.

(4) In shallow lakes, springs are likely to cause weak spots. A path should be carefully examined by chopping through the ice at frequent intervals to determine the thickness and quality, and when a safe track is found it should be marked on both sides by bushes stuck in holes in the ice.

c. Swimming.—(1) *By individuals.*—See paragraph 50.

(2) *By organizations.*—(*a*) For swimming streams of moderate width the following method is practicable and safe. A good swimmer crosses and carries with him sufficient rope or joined lariats to reach across the stream. A man on the near bank attaches the near end of the rope to the animal's halter and an equal length of rope to his tail or saddle and pushes him into the stream. The soldier on the far bank guides the animal across and the latter tows the tail rope. When the animal has reached the far bank both ropes are taken off and tied together, the knot pulled back to the near side, and an-

other animal crossed. Each man who cannot swim catches hold of an animal's tail and is towed across in safety.

(*b*) The following method is the best for crossing wide, swift rivers. Procure boats or rafts. If there are no boats and rafts must be used, then, by the use of a number of lariats fastened together and doubled, a cable may be improvised so as to make a good ferry. This cable is taken across to the far side by one or two good swimmers swimming alone or with exceptionally good animals. The cable is fastened on both banks of the stream, and the raft, which should be light but capable of holding six or eight men, is pulled across by hand, aided by poles if the stream is not too deep. The animals are stripped, except halters, and the equipment is ferried across. Several animals are crossed either by swimming or by being led from the raft. These animals, after crossing, are stood on the opposite bank in plain sight of the others. Those not yet across are held in column of twos, with the natural leaders of the herd in front. Halter shanks unsnapped from halters are fastened close around the necks, using knots that do not slip. A few men who are excellent swimmers swim with the leading animals across the stream. When the leading animals of the column see those on the opposite bank and swim toward them, the remaining animals in the column are turned loose and the whole column follows the leading animals across. It may be necessary to swim or lead across from the boat or raft quite a number of animals before the desire of the remaining animals to join them is shown. This method keeps the equipment dry and gets the timid or poor swimmers across safely.

(*c*) Streams not over 30 yards wide may be swum in column of twos without unsaddling, men retaining their seats in the saddle. Men's knees should be held well up to lessen resistance to the water; steadiness in the saddle should be maintained by grasping the pommel. This method is risky for organizations not well-trained in swimming and crossing streams with animals. Unless time is important, it is better to cross by one of the methods prescribed above. Men must be cautioned not to pull on the reins.

(3) Swimming animals must have absolute freedom of the head and neck. Absolute calmness, determination, and quiet

must prevail. The least excitement on the part of the men is sure to confuse and frighten the animals.

142. Care of animals in camp.—*a. Camp site.*—The camp site should be on well-drained high ground with shelter from cold winds or in summer with shelter from the sun. Convenience to watering places and grass must be considered. Level standings are essential.

b. Water supply.—(1) When water is obtained from a stream, the watering place for animals should be located below that designated for drinking water for men and above the place prescribed for washing clothes and bathing.

(2) The principles of watering are given in paragraph 114.

c. Restraint.—A field picket line may consist of a single long rope or cable or of several platoon picket lines. The ends should be securely pegged down or tied to wagon wheels or trees. If the line is on the ground, animals must be tied short. If the necessary materials are on hand, picket lines should be raised 3 or 4 feet from the ground. Where aerial attacks are probable, the animals must be tied on lines in groups no larger than platoons and preferably squads, and the groups should be separated from each other by at least 60 or 70 yards. Full advantage must be taken of woods, single large trees, bushes, the shadows of buildings, banks, etc.

d. Feeding.—The general principles of feeding as shown in section II should be adhered to as far as field conditions allow. See paragraph 108 *d* and *e*, concerning feed bags and feeding of hay in the field. Grazing should be afforded at every opportunity.

e. Protection of forage.—Paulins, if obtainable, should be used to cover the forage piles. The guard over animals is made responsible for the safety of all piles at the picket line. In the absence of such guards the stable sergeant is responsible to the organization commander for the protection of such forage as pertains to his organization.

f. On arrival.—(1) Animals should be walked the last 10 or 15 minutes before arrival at the camp site so that they may arrive cooled out. Dismounting and leading with loosened cinches the last 5 minutes are desirable. Immediately upon arrival in camp, after unbridling and tying on the picket line, the arms and cantle rolls are removed from saddles and the pack loads from the pack saddles. The saddles are then removed, and the blanket (without unfolding) is turned over and replaced on the back so that the dry side of the blanket is next to the back. Secure the blanket firmly in position by the use of the surcingle. The animals are allowed to eat hay placed on the picket line while shelter tents are pitched. The blanket is then removed, the back thoroughly rubbed dry and massaged with the hands, and the lower legs rubbed. The animal is then thoroughly groomed and all injuries promptly reported. Swellings resulting from saddle, cinch, or other injury should receive first-aid treatment by the rider or driver as soon as discovered. Apply a cold wet pack made from a folded grain sack or grooming cloth soaked with cold water and held in place with a surcingle. Wet packs should be resoaked frequently with cold water. By the time grooming is completed, 45 minutes or more will have elapsed since arriving in camp and the animals may be watered. After return from watering, they are fed grain, and more hay is placed on the picket line in small amounts as needed.

(2) Blankets, saddles, pack saddles, and harness should be carefully examined and, where they have caused injury, the necessary readjustments should be made to protect the injured area and to prevent repetition of the injury on the next day's march.

SECTION V

PREVENTION AND FIRST-AID TREATMENT OF DISEASE AND INJURIES

143. General.—The treatment of disease and injuries among animals, as well as prevention of disease, is the function of the Veterinary Corps. Most diseases and injuries are preventable if all concerned are vigilant, intelligent, and untiring in the application of simple preventive measures. Frequently the development of serious disease or injury may be prevented by prompt first-aid measures and early treatment. In this section the causes, prevention, symptoms, and first-aid treatment of some of the more common diseases are briefly discussed.

144. Symptoms of disease.—The most common indications of disease are partial or complete loss of appetite; temperature of 101° F. or more; accelerated breathing; increased pulse rate; listlessness; dejected countenance; profuse sweating; stiffness; nasal discharge; cough; diarrhea; constipation; pawing; rolling; lameness; inflamed mucous membranes; unhealthy coat of hair; loss of hair; itching; or unnatural heat or swelling in any part of the body.

145. Inspection for sick animals.—The best times to inspect animals for evidence of sickness or injury are while being fed and at time of grooming. One of the first and most important symptoms of sickness is impairment of appetite. Take the temperature of animals that refuse their feed. Sick animals in a corral are inclined to stand by themselves. Move each animal around and look for lameness and nasal discharge and listen for coughs. On the march, animals are inspected after arrival in camp by the veterinarian.

146. Nursing.—The chief points to consider in nursing are—

a. Ventilation.—Allow plenty of fresh air but protect from drafts. Avoid extremes of temperature and in the field provide shelter from wind and rain. Utilize grass lots and corrals whenever possible.

b. Clothing.—The amount of clothing must be regulated by the climate. In winter woolen bandages on the legs are useful, and as many as three or four covers may be used. In summer fly sheets are extremely comforting.

c. Bedding.—A good clean bed induces an animal to rest more and produces a soft springing surface for foot cases. It should be shaken up several times daily and be kept free of urine-soaked straw.

d. Stalls.—A roomy box stall, well bedded, should be used whenever possible. Keep a bucket of water in the stall and change the water frequently.

e. Shoes.—The shoes may be removed and the feet leveled if the animal is to remain in a stall for more than a few days.

f. Exercise.—Convalescent patients should receive just as much exercise as each individual case permits. However, absolute rest is one of the very best treatments.

g. Grooming.—(1) Animals that are weak and depressed should not be worried with unnecessary grooming. Such animals should be carefully hand-rubbed at least once a day, and their eyes, nostrils, and docks should be wiped out with a sponge or soft cloth. The feet should also be cleaned.

(2) Animals that are only slightly indisposed should be groomed in the usual way.

(3) Animals with tetanus should not be cleaned at all.

h. Food.—Some sick animals retain a good appetite. The principal things to observe in their cases are that they are not overfed, that droppings are kept soft, and that they have plenty of water. Sick animals with impaired appetites require special attention. They often relish a change of diet, such as a bran mash, steamed oats, chopped alfalfa, grass, roots, and apples. Feed small amounts often; do not allow uneaten portions to remain in front of them; keep mangers and feed boxes clean; sprinkle a little sweetened water over the hay and grain.

147. Stable sergeant's veterinary kit.—For use in treating animals in the absence of veterinary service or for first-aid treatment, small quantities of medicines, bandages, and instruments may be issued to stable sergeants. The doses and uses of drugs are as follows:

a. Boric acid powder.—For external use only. May be used as supplied as a drying powder for dusting on a wound or applied under a dressing. For eye or mouth wash, dissolve two mess-kit spoonfuls (level) in one mess-kit cupful of warm boiled water.

b. Iodine and potassium iodide.—Issued in tubes for making iodine solution; for external use only. Add contents of two tubes to special bottle and fill with water. Antiseptic and disinfectant on wounds. May be painted on skin over sprains, strains, or swelling. Do not use in eye.

c. White lotion tablets.—For external use only. To be used only in solution. Make solution in proportion of three tablets dissolved in one mess-kit cupful of water. Useful on swellings or sprains, either rubbed on several times daily or as a wet dressing under a bandage. Drying lotion on irritated skin, such as "scratches." Do not apply as a wet pack over wounds.

d. Cresolis solution.—Antiseptic and disinfectant for external use only. Use only in solutions as indicated. For wetting antiseptic pack to bandage over wound, add one-half mess-kit spoonful to mess-kit cup brimful of water. For washing wounds, add one mess-kit spoonful to mess-kit cup brimful of water. For disinfecting wood, leather, cloth, metal, etc., add one and one-half mess-kit spoonfuls to mess-kit cup of water.

e. Pine tar.—For external use only. Applied over or around the edges of wounds, it is useful to repel flies. Applied to hoof as tar and oakum pack.

f. Aloin capsules.—Laxative and purgative for internal use only. One capsule acts as a laxative or mild purgative. Two capsules at one dose act as a purgative. Not more than two should be given. Administer by putting far back in the mouth with the hand or with a short piece of hose. May be given by emptying contents of capsule in one-half cupful or more of water, mix well, and give with the dose syringe.

148. Classes of diseases and injuries.—*a.* Diseases are divided into two classes; communicable and noncommunicable.

(1) Communicable diseases are diseases that are transmitted or spread from animal to animal either by direct contact or indirectly through the medium of infected stables, water troughs, corrals, stock cars, food, etc. These diseases deserve more attention than noncommunicable diseases because some are incurable, some may be transmitted to man, and all may cause great losses if not held in check. The general measures for controlling this type of disease are considered in paragraph 149.

(2) Noncommunicable diseases include all diseases that are not transmissible, either directly or indirectly, from one animal to another. Examples: Colic, azoturia, heat exhaustion, thrush, etc.

b. An injury is a break in the tissue of the body resulting from external violence or from muscular activity of the body itself. The more common injuries of animals are roughly classified as wounds, bruises, strains, sprains, and fractures.

149. Communicable diseases.—*a. Prevention and control.*—(1) *General health.*—Maintain animals in good condition, feed well, groom well, keep in clean surroundings, do not overwork, protect from undue exposure, and they will resist many forms of infection. Animals in run-down condition are very susceptible to disease.

(2) *Segregation.*—Animals should be divided into three groups; affected, suspected, and healthy. The suspected animals include all those which have been in contact with the

diseased. Attendants, watering and feeding arrangements, and all equipment should be included in the separation and should be kept separate until the outbreak is over. Once an animal is placed in the affected group it should remain there until all danger is over. To avoid possible introduction of communicable diseases all animals are quarantined at least 21 days upon arrival at any new station.

(3) *Disinfection.*—Strict cleanliness of picket lines, equipment, water troughs, feed boxes, and feed bags is a good safeguard against the spread of disease. Stables, equipment, etc., which have been in contact with diseased animals should be disinfected with cresolis solution (par. 147*d*). Articles that can be boiled may be thoroughly disinfected in this manner. (See AR 40–2095.)

(4) *Inspections.*—Inspection of all suspected and healthy animals should be carried out daily. In cases where it may serve as a guide, the temperature of suspected animals should be recorded.

(5) *Destruction of animals.*—Certain communicable diseases are incurable, and it is necessary to destroy the animal to prevent spread of the disease. To shoot an animal, stand close, with the pistol almost touching the forehead and held at right angles to the head. Aim at the center of the forehead, well above the level of the eyes, just below the place where the lowest hairs of the forelock grow.

(6) *Disposal of dead.*—All carcasses of animals dying as a result of communicable disease should be burned, if practicable; otherwise buried at depth and covered with quicklime, if available. Litter should be burned over ground where discharges from the dead have fallen.

b. Influenza.—This is a very communicable disease, also known as shipping fever, affecting chiefly the respiratory system. It is spread by both direct and indirect contact.

(1) *Cause.*—Influenza is caused by the animal's eating or breathing some of the body discharges of a diseased animal, particularly the nasal and bowel discharges.

(2) *Prevention.*—Cleanliness of surroundings and good condition of animals. Quarantine of newly acquired animals for 21 days. Segregation of diseased animals. Disinfection of stall and equipment.

(3) *Symptoms.*—The first symptoms noticed are depression, great weakness, loss of appetite, rapid breathing, hacking cough, and possibly a slight watery nasal discharge. At this stage the temperature is elevated (101.5° to 106° F.), and the mucous membrane of the eyelid will have a brick red color tinged with yellow. Later the nasal discharge becomes more profuse and usually thick and yellow. Pneumonia is often a complication. Young animals frequently develop strangles at the same time they have influenza.

(4) *Nursing and first-aid treatment.*—Isolate the diseased animal. Absolute rest and good nursing are very important. Allow plenty of sunshine and fresh air but protect the body from drafts. Keep the animal warm with blankets and leg bandages during cold weather. Induce the animal to eat, as the disease is very wasting and it is important to maintain the strength with food. Give plenty of water. Do not give cathartics.

c. Strangles and distemper.—This is a communicable disease most often seen in young animals, affecting chiefly the glands in the region of the throat. It is spread by both direct and indirect contact.

(1) *Cause.*—This disease is caused by the infections (nasal, abscess, or other body discharges of a diseased animal) coming in contact with the nasal membranes or the digestive tract.

(2) *Prevention.*—The instructions given in *b* (2) above apply to this disease. The greatest danger of spread is by infected watering and feeding utensils. One attack usually renders the animal immune.

(3) *Symptoms.*—Early symptoms are loss of appetite, increased temperature, pronounced moist cough, profuse watery nasal discharge which later becomes thick and yellow, head and neck extended stiffly, and a hot and painful swelling between the jaws. The swelling usually develops after about a week into an abscess containing very thick yellow pus. In more serious cases abscesses may develop in other parts of the body.

(4) *Nursing and first-aid treatment.*—Segregate sick animals and see that they have absolute rest. Clothe the body according to the weather and prevent drafts but provide plenty of fresh air. Paint the swelling between the jaws with tinc-

ture of iodine once daily. Tempt the appetite and provide food that is easily chewed. Clean the discharge from the nostrils two or three times daily with cresolis solution of strength for washing wounds, and after abscess opens clean two or three times daily with same solution. Open abscess with knife when it is well "pointed" and nearly ready to burst (indicated by loosening of hair, softening and moistening of skin at the point).

d. Pneumonia.—Pneumonia is an inflammation of the lung tissue. It is frequently a complication of influenza. It should be handled as a communicable disease.

(1) *Causes.*—A complication of other communicable diseases of the respiratory system, especially of influenza; overexposure to cold, especially when heated; overexertion; or irritating fluids or gases entering the lungs.

(2) *Symptoms.*—A severe chill, with a temperature varying from 103° to 107° F.; redness of the mucous membranes; rapid, difficult breathing; and a full rapid pulse (from 50 to 80 per minute). The nostrils are dilated and the expired air is quite warm. There is frequently a reddish discharge from the nose and there may or may not be a cough. The animal is usually constipated at first, and the urine is scanty and high colored. The legs and ears are cold and there is great weakness.

(3) *Nursing and first-aid treatment.*—Isolate the animal in a clean, dry place free from drafts, but abundantly supplied with fresh air. Clothe the body according to the season; rub the legs well and bandage warmly. Remove the bandages twice daily, rub the legs well, and reapply. Feed easily digested food (bran mashes, grass, good hay, and steamed oats), and keep a supply of fresh water within reach at all times.

e. Coughs and colds.—Animals suffer from coughs and colds quite similar to the same conditions in man. They are mildly infectious inflammations of the membranes of the nose and throat.

(1) *Causes.*—Exposure to wet or cold, particularly when tired and heated; damp or poorly ventilated stables; sudden changes in weather.

(2) *Symptoms.*—Dryness and redness of the nasal membranes followed by a watery discharge which in a day or so becomes grayish and thickened, later yellow in color. If the

throat is affected, the cough is at first dry and later moist. The nasal discharge is odorless. In the early stages the animal is somewhat listless and may run a slight temperature, 101° to 102° F.

(3) *Nursing and first-aid treatment.*—From early symptoms it is difficult to tell whether the condition is a simple cold or the beginning of influenza or possibly strangles. For this reason the animal should be handled as if he had influenza (par. 148*b*). Simple colds will usually respond to rest and a mild, laxative diet. The animal must be protected from cold and drafts or overheating.

f. Glanders.—Glanders is a chronic, incurable, and very communicable disease affecting horses and mules. Man is also susceptible.

(1) *Cause.*—The organism that causes the disease is found in the nasal discharges, bowel discharges, and secretions from ulcers on the skin. It may enter the body through the respiratory system, digestive system, or surface wounds.

(2) *Prevention.*—The most effective weapon in preventing glanders is the mallein test. This test detects infected animals in the early stages of the disease before they have become spreaders of infection, or before they have shown any external signs of disease. The infected animal is at once destroyed. Military animals should not drink at public troughs or occupy private or public stables not previously inspected by a veterinary officer. Upon the discovery of a diseased animal in any organization, stringent quarantines, disinfection, testing, and other control regulations are at once put in effect.

g. Tetanus.—Tetanus is a disease which affects the nervous system.

(1) *Cause.*—The germ that causes tetanus is found in the ground and gains entrance to the body through wounds. Nail punctures of the hoof or deep wounds of the legs are most likely to be infected.

(2) *Prevention.*—This disease is entirely preventable by giving the animal a dose of tetanus antitoxin immediately after the wound is inflicted.

(3) *Symptoms.*—Stiffness and spasms of some or all of the muscles usually develop within 5 days after infection. The first symptom is muscular stiffness interfering with movement,

mastication, swallowing, and drinking. As the disease progresses the gait becomes stilted, with little bending of the joints and the head is carried stiffly extended. There is little increase in temperature in the early stages.

(4) *Nursing and first-aid treatment.*—Isolate the animal in a quiet, darkened stall where he will be free from anything that may excite him. All wounds should be thoroughly disinfected with tincture of iodine. Feed gruels of thin mashes and keep water in front of the animal at all times. Antitetanus serum is of little value after symptoms have developed. Diseased animals that live as long as 10 days usually recover.

h. Equine encephalomyelitis.—Equine encephalomyelitis is an acute infectious disease of horses and mules which affects the central nervous system.

(1) *Cause.*—The disease is caused by a filtrable virus. The infective virus is carried from the sick to well animals only through the medium of infected mosquitoes. A few days after biting a diseased animal, the mosquito becomes infective and by its bite may transmit the disease to well animals.

(2) *Prevention.*—Avoid areas of infection. Use all possible means to protect animals from mosquitoes. At the present time it appears that annual vaccination with a newly developed type of vaccine will result in a high percentage of protection against the infection.

(3) *Symptoms.*—The animal usually shows symptoms of drowsiness. Excitement is sometimes shown in the early stages. The gait is staggering and the animal frequently walks aimlessly in circles or may stand quietly with the head pressed against a stall or fence. Often the animal can swallow only with difficulty or not at all, and saliva mixed with partially chewed food may drool from the mouth and nose. Later the animal may fall to the ground and be unable to rise.

(4) *Nursing and first-aid treatment.*—Immediate isolation of suspected and diseased animals with complete protection from mosquitoes. Encourage eating and supply plenty of clean drinking water. Endeavor to keep the animal standing. Cold packs applied to the poll and frequently changed are of value.

i. Mange.—Mange is a communicable parasitic skin disease spread by both direct and indirect contact.

134205°—39——8

(1) *Cause.*—A very small animal parasite or mite which lives on or in the skin. It may be spread by immediate contact with diseased animals or by infected equipment, stables, cars, etc.

(2) *Prevention.*—Immediate segregation of diseased animals and disinfection of all stables and equipment that may have become contaminated. Avoid interchanging of equipment among animals. Use individual grooming equipment.

(3) *Symptoms.*—Patchy loss of hair associated with intense itching. The surface of the affected areas is likely to be moist and raw or scabby, due to the activity of the parasite and scratching or rubbing by the animal. Later the skin becomes much thickened and tends to lie in ridges.

(4) *Nursing and first-aid treatment.*—Segregate diseased animals. Clip and burn hair. For small areas, scrubbing the skin with cresolis of strength recommended for washing wounds may be of some benefit. Routine hand treating or dipping with lime and sulphur or arsenical dip is the usual method of treatment employed.

j. Ringworm.—Ringworm is a communicable skin disease spread by direct and indirect contact.

(1) *Cause.*—Caused by a fungus quite similar to ordinary mold. The causative fungus will remain infected in stables, corrals, and on equipment for long periods.

(2) *Prevention.*—Good grooming. Use only clean equipment. Avoid interchange of grooming and other equipment in contact with animals. Segregate infected animals. Disinfect all horse equipment, stables, etc. Clipped animals are less frequently infected.

(3) *Symptoms.*—On the affected patch, the hair becomes erect and dull. After a few days, this hair, with a dry scab, falls off leaving a bare, circular patch varying in size from a penny to a dollar.

(4) *Nursing and first-aid treatment.*—Segregate diseased animal and disinfect equipment. With a brush scrub the affected area with cresolis solution of the strength of one mess-kit spoonful to a mess-kit cup brimful of water. Loosen and remove scabs. Thereafter apply tincture of iodine about every second day after rubbing off the loose scabs or scales.

k. Lice.—A small parasite that lives on the skin.

(1) *Cause.*—Spread by direct contact between animals and by infested stables or equipment. Usually contracted from stables which may remain infested from year to year if not disinfected.

(2) *Prevention.*—Good grooming, clipping, isolation of infested animals, and disinfection of stables and equipment.

(3) *Symptoms.*—Scratching, loss of hair due to rubbing, presence of lice, presence of small white eggs (nits) stuck to shafts of some of the hairs. Neck, mane, and back are regions most frequently infested.

(4) *Nursing and first-aid treatment.*—Segregate infested animals. Clip the animal and burn the hair. Wet the entire body with cresolis in the strength solution of one mess-kit spoonful of cresolis to a mess-kit cup brimful of water. Bathe the body again in this solution at the end of 5 days. Disinfect stables and equipment.

150. Noncommunicable diseases.—*a. Heat exhaustion; overheating; sunstroke.*—These are noncommunicable disturbances of the nervous system due to heat.

(1) *Cause.*—Caused by long continued hard or fast work during hot or very humid weather, especially among animals not in good condition or having heavy coats.

(2) *Prevention.*—Do not overtax the strength of the animal. Watch animals for early symptoms. Clip animals that have heavy coats. Water frequently on hot days.

(3) *Symptoms.*—Thumps, a condition described in *b* below, often precedes overheating. The animal that has been sweating freely will cease to sweat and will be dull and require urging. The gait is staggering or wobbly, especially in the hindquarters. If halted, the animal stands with the legs spraddled; breathing very rapid and shallow; nostrils dilated; expression drawn and anxious; nasal membranes bluish red in color; and trembling of body muscles. The body feels hot to the hand and the temperature will be from 103° to 109° F.

(4) *Nursing and first-aid treatment.*—Prompt first-aid treatment is of utmost importance. Stop the animal at once and in the shade if any is nearby. Remove the equipment and apply large quantities of cold water to all parts of the body but especially to the head, sides of the neck, groins, and flanks. Inject cold water into the rectum with a syringe. Wash out

the mouth and nostrils with cold water. Give the animal three or four swallows of water every few minutes. Under this treatment the temperature will drop quite rapidly, and as improvement is noted move the animal about very slowly and rub the body to prevent chilling. As soon as the temperature is near normal the animal may be moved slowly on into camp.

b. Thumps.—Thumps are spasms of the diaphragm.

(1) *Cause.*—Overwork or fast work during hot weather especially among animals not properly conditioned.

(2) *Symptoms.*—General symptoms of fatigue with spasmodic jerking noticeable in the belly and flanks; frequently a distinct thumping sound will be heard.

(3) *Nursing and first-aid treatment.*—If marching, halt the animal and if the temperature is elevated reduce it by sponging the body with cold water and then have the animal ridden or led into camp at the walk.

c. Exhaustion.—(1) *Causes.*—Overexertion; excessive or prolonged heavy work; lack of condition.

(2) *Symptoms.*—After the animal arrives in camp, he may lie down and refuse his feed, especially his grain, yet drink considerable quantities of water. The temperature may be slightly elevated and the pulse may be weak and thready. Sweating may be quite noticeable and possibly patchy, yet the body feel cold and clammy.

(3) *Nursing and first-aid treatment.*—Make a comfortable place for the animal to lie. Cover the body with a blanket to prevent chilling. Hand rub the legs. Give small amounts of water frequently. A period of rest is all that is needed to recuperate from excessive fatigue.

d. Colic.—A general term applied to abdominal pain caused by digestive disturbance either spasmodic and flatulent or gas colic.

(1) *Causes.*—The causes of both forms of colic are very similar, but gas colic is more frequently caused by foods fermenting in the digestive tract. General causes are indigestible or spoiled food, sudden changes in food, overeating, eating while fatigued, working too soon after feeding, watering while exhausted or hot, bolting the feed, overeating of green feed, and watering too soon after feeding. Windsucking is fre-

quently a cause of gas colic. Collections of sand in the bowel may result in repeated attacks of colic.

(2) *Prevention.*—Close attention to the principles of feeding and watering will prevent most cases of colic.

(3) *Symptoms.*—Pain as indicated by restlessness, pawing, stamping of the feet, looking around at the flanks, kicking at the abdomen, lying down, rolling, sweating, and frequent attempts to defecate usually resulting in the passage of but a few pellets of dung or a discharge of gas. In spasmodic form, the attacks are often intermittent with short periods of a few minutes of apparent freedom from pain. In the gas type of colic, the digestive tract is filled with gas, the belly is distended, and breathing is difficult.

(4) *Nursing and first-aid treatment.*—Place the animal in a well-bedded box stall or, if in camp, bed down a section of the picket line and have an attendant hold the animal's tie rope. Get a veterinarian at once, if possible. Do not attempt to keep the animal from rolling, etc., unless he is throwing himself to the ground so violently that it is evident he may rupture some organ. Give the animal two aloin capsules. Give frequent rectal injections of 2 or 3 gallons of warm soapy water. Wring blankets out of hot water and wrap around the belly and flanks as hot as can be borne without burning the hands or animal. Water may be given in small amounts. Withhold all food until at least 12 hours after all pain has disappeared and then feed light for 2 or 3 days.

e. Diarrhea.—(1) *Causes.*—Spoiled food, overfeeding of "washy" feeds, sudden changes of diet; and nervousness.

(2) *Prevention.*—Careful attention to kind, quality, and quantity of feed and methods of feeding. Exclude "washy" feeds from the diet of animals which tend to scour.

(3) *Symptoms.*—The droppings are frequent and of semifluid nature. If the condition continues long, the animal loses flesh and appetite is wanting.

(4) *Nursing and first-aid treatment.*—Correction of diet in mild cases will be sufficient. Give one aloin capsule and give the animal absolute rest while withholding all food and limiting the amount of water for a period of 24 to 48 hours.

f. Azoturia.—(1) *Cause.*—Caused by violent exercise following idleness.

(2) *Prevention.*—When conditioned animals accustomed to regular work are given a period of complete rest for longer than a day, reduce the grain ration by at least one-half. When animals are exercised after a period of rest, they should be walked for at least 20 minutes after leaving the stables and not called upon to do more than a very small amount of fast work the first day.

(3) *Symptoms.*—Increased excitability, profuse sweating, and rapid breathing are the first symptoms. Very soon the animal begins to stiffen in his hindquarters, drag the hind legs and knuckle over in the hind fetlocks. If continued in work the animal will become completely incapable of supporting weight on the hind legs and fall to the ground, and in such cases the chances of recovery are remote. The urine is scanty and red or coffee-colored.

(4) *Nursing and first-aid treatment.*—Stop the animal immediately when the first symptoms are observed. Remove the saddle or harness and cover with three or four blankets. Keep the animal standing, if possible; if not, provide a good bed. Heat some oats or common salt, place in a sack and spread over the loins to relieve the pain. If hot water is available, a hot blanket wrung out, placed over the back and loins, and covered with dry blankets is very beneficial. After a few hours the average case can be moved slowly to the stable, provided the distance is not too great. At this time he should be given a purgative (aloin capsule) and be fed on bran mashes, grass, and hay for a few days.

g. Laminitis (founder).—(1) *Causes.*—Overeating grain, eating improper or spoiled food, colic, exhaustion, overexertion, long continued work on hard-surfaced roads, and drinking cold water while overheated.

(2) *Symptoms.*—Intense lameness, which appears quickly. In mild cases the animal moves stiffly, taking short, rapid steps with the forefeet and with the hind legs carried well forward under the body to relieve the forefeet from the weight of the body. Usually only the forefeet are affected. The affected feet are very hot. The pulse and respiration are greatly accelerated and temperature may reach 105° F.

(3) *Nursing and first-aid treatment.*—Remove the shoes from the affected feet and place the animal in a well-bedded

box stall. Cover the affected feet and legs as high as the knees and hocks with several layers of burlap and keep saturated with cold water. If running water is available, place the animal in it and allow him to stand for 24 to 36 hours. In this case the shoes need not be removed. Laxative diet is indicated. Give one or two aloin capsules. Feed bran mashes and a little hay. Early treatment is an important factor. Cases that are not cured in 4 or 5 days are likely to develop into chronic laminitis which is practically incurable. As soon as the acute pain has diminished, moderate walking exercise each day is beneficial. Following the attack, shoe with a bar shoe over a leather pad covering a tar and oakum pack.

h. Scratches.—An inflammation in the back of the pastern region.

(1) *Causes.*—Wet, muddy, and filthy standings; failure to dry legs that have become wet from slush, rain, or washing; pasterns not thoroughly cleaned while grooming; short clipping of the hair on the back of the pastern. Most prevalent during wet, cold weather.

(2) *Symptoms.*—Redness, heat, pain, and swelling of the skin on the backs of the pastern. Later the surface of the skin becomes moist and raw, and dust and dirt dry with the secretions to form a scabby mass sticking to the skin and hair. The skin may crack. Usually there is lameness.

(3) *Nursing and first-aid treatment.*—With soap and warm water carefully soak off all accumulated dirt and dried secretion. Rinse with clean, warm water and dry. Apply a white lotion pack under a bandage twice daily. Rest the animal on clean, dry standings and feed laxative foods. After the moistness of the skin has decreased, apply dry powdered boric acid held in place by cotton in a loose bandage. Avoid the use of water after the first cleansing.

i. Acute lymphangitis.—(1) *Cause.*—Failure properly to reduce grain ration of a heavily fed, well-conditioned animal during a period of idleness.

(2) *Prevention.*—Reduce grain ration of working animals at least one-half during periods of enforced idleness.

(3) *Symptoms.*—Marked swelling, heat, tenderness, and lameness in one or more legs, usually but one hind leg. The tenderness is most pronounced on the inside of the leg between

the hock and body. Loss of appetite; rapid pulse and respiration; and a temperature of from 102° to 106° F. The leg may swell to twice its normal size, and in some cases the skin will become moist from secretions and will crack open.

(4) *Nursing and first-aid treatment.*—Give two aloin capsules. Feed animal hay only. Allow plenty of water. Shower leg with cold water continuously during the day or bathe with cold water for 20 minutes several times daily. After each showering or bathing of the leg apply white lotion solution. After pain has diminished give slow walking exercise followed immediately by warm bathing and hand rubbing of the leg.

j. Thrush.—(1) *Causes.*—Failure to clean out the depths of the commissures and cleft of the frog, lack of frog pressure, filthy standings, dryness of the feet, and cuts or tears in the horny frog are all contributing causes.

(2) *Prevention.*—A hoof that is properly groomed once each day will not develop thrush. A thorough washing of the under surface of the hoof once a week will materially assist in prevention of this disease.

(3) *Symptoms.*—Cracks, depressions, or fissures in the horn of the frog in which is found a thick, dark colored discharge with a very offensive odor. The cleft of the frog and the sides of the frog at the depths of the commissures are the parts usually diseased.

(4) *Nursing and first-aid treatment.*—Clean and wash the hoof. With a sharp hoof knife, trim away all diseased and under-run horn and all ragged pieces. With cresolis solution (one and one-half mess-kit spoonfuls to mess-kit cupful of water) and a stiff brush thoroughly scrub the horn. After it has dried, paint the area with iodine. Repeat the washing and iodine treatment daily until the horn begins to appear dry and then apply pine tar.

k. Tendinitis.—This is an inflammation of the large tendons on the back of the leg in the cannon region.

(1) *Causes.*—A strain of these tendons; long toes and low heels; violent efforts and sudden checks, as in jumping or galloping over rocky, uneven ground; long-continued exertion in which the muscles tire and are more easily strained; lack of fit condition; tight bandaging.

(2) *Prevention.*—Proper balance and shoeing of the feet. Have the animal in good, fit condition. Avoid other causes of the injury.

(3) *Symptoms.*—Lameness; heat and swelling in the tendon; sensitiveness of the tendon to pressure. The tendons of the forelegs are far more frequently affected than are those of the hind legs. The tendons in the cannon region may be affected their entire length or only in a part of their length.

(4) *Nursing and first-aid treatment.*—Absolute rest is most essential. Shower the leg with cold water from a hose or apply cold or ice water packs throughout the day, and for the night apply a white lotion pack over the injured tendon. After the acute swelling and tenderness have disappeared, bathe with hot water and massage, then rub the tendon briskly with tincture of iodine once daily.

151. Injuries.—*a. Rope burn.*—This is an injury usually occurring on the back of the pastern.

(1) *Causes.*—Tying with the tie rope too long is the most frequent cause. Often caused by the animal getting his foot, usually a hind foot, over his own or an adjacent horse's tie rope or entangled in a picket line guy rope. An improperly made or improperly used side line or casting rope often results in rope burns.

(2) *Symptoms.*—A simple chafe or abrasion of the skin. It may involve the underlying tendon. Lameness is usually a symptom.

(3) *Nursing and first-aid treatment.*—General methods of treatment same as for scratches (par. 150*h*) except that white lotion packs should not be used if the injury is more than a surface one.

b. Lameness.—(1) Lameness may best be detected at the trot. When an animal is lame it takes as much weight as possible off the injured leg and places it on the opposite one. If lame in a foreleg, the animal will be seen to nod its head every time the sound foot comes to the ground. If lame behind, the hock of the sound leg comes higher and dips lower than that of the lame one, and the head may nod as the lame foot strikes the ground. Animals lame in both fore and hind legs take short strides with both; they idle along instead of striding out.

(2) The majority of causes of lameness occur at or below the knee and hock. The foot is the most common seat. Lameness in the shoulder is comparatively rare. In all cases where no apparent or sufficient cause can be detected the foot should be thoroughly examined.

c. Fractures.—A fracture is a broken bone.

(1) *Causes.*—Most fractures result from kicks inflicted by other animals. The bone forming the point of the hip may be fractured by falling on the side or by striking it against the side of a door.

(2) *Symptoms.*—When any long supporting bone of the leg is completely fractured, the leg dangles helplessly and will bear no weight. When the bone forming the point of the hip is fractured the animal may show but few symptoms other than lameness, difficulty in advancing the hind leg on that side, tenderness and swelling over the seat of injury, and a noticeable lowering of the point of the hip on that side.

(3) *Nursing and first-aid treatment.*—Complete fractures of any of the supporting bones of the legs of horses or mules are considered incurable, and the destruction of the animal is authorized without delay. Destroy by shooting as described in paragraph 149*a* (5). Fracture of the bone forming the point of the hip will heal provided the animal is given complete rest for a month or six weeks.

d. Sprains.—A sprain is a joint injury usually without any break or injury of the overlying skin.

(1) *Causes.*—Twisting or pulling of the joint or bending the joint beyond its normal range of action resulting in stretching or tearing of the ligaments.

(2) *Symptoms.*—Marked lameness; heat and swelling over the joint; tenderness on pressure or manipulation.

(3) *Nursing and first-aid treatment.*—Treatment should be the same as that outlined for tendinitis (par. 150*k* (4)).

e. Wounds.—(1) *Classification.*—Wounds are divided into—

(*a*) Incised wounds or cuts.

(*b*) Lacerated wounds or tears.

(*c*) Punctured wounds or holes.

(2) *Treatment.*—(*a*) *Cleanliness.*—Cleanliness of the wound itself, of the dressings, and the dresser is of the greatest importance.

(*b*) *To stop bleeding.*—Stoppage of bleeding is the first point to be attended to. Tight bandaging above and below the wound or a pad on the wound is usually sufficient to control the flow. If a large blood vessel is cut and the end exposed, it should be tied around with clean thread which has been soaked in antiseptic.

(*c*) *To clean.*—Cut the hair from the edges of the wound and remove all dirt, clots of blood, splinters, and foreign bodies of all kinds. This may be done by carefully syringing the parts with clean warm water or a warm antiseptic solution.

(*d*) *To close.*—Sutures and bandages are used for this purpose, but no wound that has been dirty must ever be entirely closed. Sutures may be used in parts where there is little flesh, such as around the forehead, eyelids, and nose. They are less useful in the fleshy parts, because the movements of the muscles and swelling cause them to pull out. In applying sutures the borders of the wound must be brought together in their natural position. The needle and thread are passed through the skin at one side of the wound and out at the other. The sutures should be from ¼ to ½ inch from the edge and about ¾ of an inch apart. They should be drawn just tight enough to bring the edges of the skin together. As a rule, they should be removed in about 8 days.

(*e*) *Drainage.*—Drainage at the lowest part of the wound is necessary for the removal of pus. In horizontal wounds a small vertical opening must be made below the line of the stitches.

(*f*) *Dressings.*—Wounds should be dried carefully with gauze or cotton, treated with an antiseptic, covered with dry gauze or cotton, and a bandage applied; or cotton soaked in antiseptic may be put on and held in position by a bandage, care being taken to avoid undue pressure. If the location will not permit bandaging, the injured parts may be treated with an antiseptic and covered with a clean piece of cloth or gunny sack, the inside of which may be lined with a piece of gauze large enough to cover the wound. Bandaging for too long a period may prevent the area from covering itself with skin, resulting in proud flesh. After a wound has begun to granulate (fill in with repair tissue), it will often heal much better if no dressing or bandage is placed in contact with the wound, provided it is not irritated by flies.

(*g*) *Rest and restraint.*—If the injury is slight the animal may continue at work; otherwise he may be kept in a box stall, cross-tied, or subjected to some other form of restraint.

(*h*) *After care.*—All wounds should be kept dry and dressings should be changed only often enough to keep the wound clean. As little washing as possible should be done, and the parts should be sopped instead of rubbed.

(*i*) *Flies.*—The healing of wounds that cannot be covered is sometimes retarded by the presence of flies. The edges of such wounds, and also the surface if not too moist, may be covered lightly with pine tar.

(*j*) *Maggots and screwworms.*—Wounds sometimes get flyblown and maggots appear. Their presence is recognized by a thin, bloody discharge from the wound and the red, angry appearance of its edges. If the bottom of the wound is carefully examined, movement of the worms may be seen. With forceps pick out all the worms and wipe out the cavity with a swab of cotton that has been saturated with tincture of iodine.

(*k*) *Excessive granulations (proud flesh).*—In sluggish, slow-healing wounds small, rounded fleshy masses which protrude beyond the edges of the wound are often formed. These fleshy masses are called proud flesh. The growth may be kept down by removing with scissors to the level of the skin and treating with boric acid or tincture of iodine, or by searing to the level of the skin with a heated iron, care being taken to see that the surrounding skin is not burned.

f. Contusions (bruises).—A contusion is an injury of a part without breaking the overlying skin.

(1) *Causes.*—Blow from a blunt object. Being kicked is a common cause.

(2) *Symptoms.*—When over a muscle, they produce soreness and lameness and, if severe, may rupture a blood vessel which will result in a fluctuating swelling caused by an accumulation of blood usually just under the skin. Such injuries are most common on the thigh and buttocks. Contusions over a joint may be mistaken for a sprain of the joint.

(3) *Nursing and first-aid treatment.*—Rest and general method of treatment as outlined for tendinitis (par. 150 *k* (4)).

In bruises where a fluctuating swelling results that does not reduce in a week or ten days, the swelling may be opened with a knife at its lowest point to allow the fluid contents to escape. After opening, the external parts should be kept clean and the external wound painted with iodine once daily.

g. Summer sores.—(1) *Cause.*—Apparently due to the effects of irritation of an ordinary wound by flies. They invariably occur during hot weather when flies are most prevalent.

(2) *Prevention.*—Protect all wounds from flies by appropriate dressings or medicinal application.

(3) *Symptoms.*—The wound tends to form proud flesh rapidly, and the surface of the area is porous and exudes a considerable amount of thin discharge. Near the surface of the proud flesh and in its depths may often be found small, hard grayish or yellowish bodies having much the appearance of cracked corn.

(4) *Nursing and first-aid treatment.*—Remove pronounced growths of proud flesh. Thereafter treat as other wounds with special protection from flies.

h. Eye injuries.—(1) *Causes.*—Eye injuries are usually due to blows, scratches, or punctures from brush, forage, etc., and to small objects lodging on the front of the eyeball or under the lids.

(2) *Symptoms.*—Watery eye, flow of tears from the eye, reddened membranes, and partial or complete closing of the eye are symptoms of all forms of eye injury. If the cornea or clear part of the eyeball itself is injured, the cut or wound may be visible, and surrounding it the eye will become milky white in color. Ordinarily foreign matter merely lodged behind the lids does not cause the eyeball to become milky in color.

(3) *Nursing and first-aid treatment.*—If there is any foreign body under the lids, flood it out with clean water using a syringe, or remove by the careful use of a tightly rolled swab of cotton or gauze. Then flood the eye several times daily with boric acid solution prepared by dissolving two level mess-kit spoonfuls of boric acid in a mess-kit cupful of warm water. Use an eye dropper for this purpose. Cover the eye with a pad of cotton about 6 inches square covered with gauze, and hold in position with strings tied about the head and to the halter, or sew the pad on the inside of an improvised head

bandage made from a grain sack. In bad cases soak the pad in boric acid solution before applying. Keep the animal in a dark place.

i. Treads and overreaching.—(1) *Causes.*—Treads are injuries or wounds in the coronet on the front or sides of the foot which may be self-inflicted or inflicted by the shoes of other animals. They are caused frequently by animals crowding in stables, through doors, and in shipment. Overreaching wounds are self-inflicted and are injuries to the bulbs of heels, pastern, or fetlock, caused by the animal's hind foot striking the part. These injuries are most frequently inflicted when landing after jumping an obstacle, at the extended trot, and at times at the gallop. Wounds similar to overreaching wounds are often caused on the heels of the hind feet by another horse crowding up on the animal from the rear.

(2) *Prevention.*—The nature of some causes suggests the means of prevention. Shoeing with rocker-toed shoes and high heels which speed up the action of the forefeet is often beneficial. Jumping horses, in which the injury cannot be prevented otherwise, should be jumped with bell boots.

(3) *Symptoms.*—Tread wounds usually are in the coronet and the skin, and often some of the horn at the top of the hoof is torn. There is usually heat and swelling in the part. Most overreaching wounds occur at the bulbs of the heel; the skin above the horn is more bruised and scraped than cut; and the horn at the heel is torn loose and may be separated from the sensitive tissues, or the wound becomes infected.

(4) *Nursing and first-aid treatment.*—For either injury clip the hair around the wound and clean thoroughly with cresolis solution, removing all foreign material. With scissors cut off all loose flaps of skin or tissue and, with a hoof knife or rasp, thin somewhat the horn in the region of the injury. Saturate a small pledget of cotton with iodine and bandage tightly over the injury. Dress in this manner daily. When the wound appears dry and is healing, cover with tar and a small pad of oakum and bandage tightly.

j. Interfering wounds.—Interfering wounds are self-inflicted wounds made on some part of the inside of the leg, usually the fetlock, by being struck by the hoof or shoe of the opposite leg.

(1) *Causes.*—Defective conformation, such as toeing out, cow-hocked, and narrow breasted; defective shoeing; and traveling over uneven footing. Fatigued animals are more likely to interfere behind.

(2) *Prevention.*—Corrective shoeing, etc. If not correctable, use leather or felt interfering boots, or pad the part with oakum held in place with a bandage.

(3) *Symptoms.*—The injury may vary from a slight roughing of the hair at the coronet or inside of the fetlock to a deep wound causing marked lameness. At the moment of interfering an animal will often carry the leg for a few steps without putting weight on it, then go very lame for a few steps and soon proceed without lameness.

(4) *Nursing and first-aid treatment.*—Treat the same as other wounds (par. 151*e* (2)). Apply a thick, padded dressing to avoid further injury.

k. Corns.—A corn is a bruise or an inflammation of the sensitive tissues underlying the horn in the region of the heels.

(1) *Causes.*—Caused by external bruising such as by overreaching or by bruising due to concussion within the foot incident to the foot striking the ground while the animal is in movement. Fast or long-continued work on hard-surfaced roads which increases the normal concussion may cause corns or general foot soreness. Lack of frog pressure, dry feet, high or unduly low heels, leaving shoes on too long, and contracted heels are all possible causes of corns.

(2) *Prevention.*—Shoe for maximum frog pressure, avoid loss of horny frog by thrush, maintain the normal moisture content and elasticity of the horn of the hoof, avoid unnecessary hard or fast work on unyielding surfaces, prevent contraction of the heels by proper hoof care and corrective shoeing, and avoid the unnecessary use of calked shoes.

(3) *Symptoms.*—Lameness is a symptom. With mild corns in both forefeet, lameness may be evidenced only by a slight stiffness and shortness of gait in the forelegs. One or both forefeet may be affected or only one heel of a foot. The affected heel and quarter may be hot, and the animal will flinch when pressure is exerted on the affected area with the hoof testers or tongs.

(4) *Nursing and first-aid treatment.*—Remove the shoe. Reduce the inflammation by cold packs on the hoof or by standing in cold water. Lower the wall at the buttress and quarter to relieve that part from bearing on the shoe. Shoe with a bar shoe and a pad to get increased frog pressure. If pus has formed, the under run horn should be removed and free drainage established. General treatment is used for wounds until dry, then tar and oakum dressings and corrective shoeing.

l. Penetrating wounds of the foot (nail pricks, etc.).—(1) *Causes.*—Most penetrating wounds of the foot are caused by the animal stepping on nails or screws or by a misdirected shoeing nail.

(2) *Nursing and first-aid treatment.*—If the nail is still imbedded in the horn, before removing it, clean all dirt from the bottom of the foot by washing and then wash with cresolis solution. Remove the nail or other object, and with a knife thin the horn over a fair-sized area surrounding the penetration and make a final opening about ⅛ to ¼ inch in diameter through the horn to the sensitive tissue. Saturate a rather small pledget of cotton with tincture of iodine and place it on the wound. Over this place a pad of oakum covering part or all of the bottom of the hoof and hold in place with strips of tin or light sheet metal of such length and width that their ends can be engaged between the hoof and the shoe on the inside at both sides of the shoe. Do not probe the wound in the sensitive tissue and do not compress the dressing so tightly that the secretions are dammed back into the wound. Dress daily with iodine until the wound appears to be dry and healing and then apply a tar dressing under an oakum pack about every 3 days until no further dressing is necessary.

m. Dry feet.—(1) *Causes.*—Lack of frog pressure, lack of exercise, dry weather, allowing the shoes to remain on the feet too long, and loss of the wax-like horny covering (periople) of the horny wall.

(2) *Prevention.*—Make provision for frog pressure and exercise the animal regularly. During dry weather pack the bottom of the hoof with wet clay.

(3) *Symptoms.*—The first symptom is the hardness and drying of the horny frog. While dryness in itself may not cause

lameness, it is an active cause of contracted heels, corns, thrush, etc.

(4) *Nursing and first-aid treatment.*—The first step should be to correct errors in shoeing and the second to restore and maintain the normal moisture content of the horn. Moisture may be restored to the horn by packing the feet daily with wet clay, by standing the animal in a shallow clay mud bath, by wet packs on the feet, or by standing in a foot bath. The application of some vegetable or animal oil or pine tar to the wall of the hoof will assist in preventing the loss of moisture by evaporation.

n. Contraction of quarters and heels.—(1) *Causes.*—Dry feet, thrush, cutting of the bars, cutting away the horny frog, continued use of heel calks, and allowing the shoes to remain on the feet too long.

(2) *Symptoms.*—The quarters, and especially the heels, are narrow and high and there is little or no frog pressure. The frog atrophies and is often affected with thrush.

(3) *Treatment.*—(*a*) Soften the feet as prescribed for dry feet.

(*b*) If the contraction is not pronounced, and the horny frog is in a healthy condition and of sufficient size to come in contact with a bar shoe, use the bar shoe well rolled at the toe, pack the feet with oakum and tar, and cover with a leather pad.

(*c*) If the contraction is pronounced the following method may be used: Remove the wall proper from the affected quarter or quarters, if both are contracted, from a point near the bend of widest part of the hoof, back to and including the buttress and from the coronary band to the lower border of the hoof. The wall proper should be removed with the rasp and knife being careful not to draw blood. Shoe with a bar shoe and pack the foot with pine tar and oakum. Dress the cut surface of the wall with pine tar.

o. Quarter crack; toe crack.—(1) *Causes.*—Dryness of feet, lack of frog pressure, contracted heels, injuries or deformity of the coronary band, long toes, large nails, and nails driven too near the heel.

(2) *Symptoms.*—A crack in the wall of the hoof which begins at the coronary band and extends downward. Only those

134205°—39——9

cracks which extend through the entire thickness of the wall and cause bleeding or lameness are of consequence.

(3) *Nursing and first-aid treatment.*—Soften the horn of the wall by wet packs, bran poultices, or standing the animal in water for a few days. After this, clean out the crack and cut away the overlapping edges of the horn. For an inch or more on each side of the crack and for 1½ inches or more downward from the coronary band, rasp the wall as thin as possible without injuring the "quick" or drawing blood. For a quarter crack trim away the bearing surface of the wall from a point ½ inch in front of the crack back to and including the buttress so that the affected quarter does not come in contact with the shoe. Shoe with a bar shoe, with good frog pressure, over a leather pad with tar and oakum. Keep the wall soft with daily application of tar, especially over the crack. A little tar rubbed vigorously into the coronary band once a week will stimulate the growth of horn.

p. Sore backs.—Sore backs are probably the greatest cause of disability among cavalry mounts. Sore backs can be prevented.

(1) *Causes.*—Improper saddling, poorly fitted saddles, dirty or improperly folded saddle blankets, careless riding such as lounging in the saddle, unevenly placed loads, and poor conformation.

(2) *Prevention.*—(*a*) The individual rider must be held strictly responsible that the adjustment of his saddle, arms, and pack is correct; that his blanket is clean and accurately folded; and that he reports at once to his immediate superior the slightest injury of any kind discovered on the back of his animal. Officers and noncommissioned officers must prevent men from slouching in the saddle. Riders must sit erect in the saddle at all times. Riding on the cantle or standing in one stirrup is sure to bring saddle sores.

(*b*) Intelligent application of the principles of correct saddling can usually overcome poor conformation or poorly fitted saddles. Small pads made from old blankets or strips of felt tacked on the bars take the place of lost flesh.

(3) *Treatment.*—(*a*) Ascertain and remove the cause. In fresh cases where the skin is not broken, apply cold irrigations or baths with gentle hand rubbing. This should be followed

by the application of packs saturated and kept wet with cold water and held in position by means of a surcingle or bandage.

(*b*) Injuries to the withers and ridge of the spine should be irrigated or bathed with cold water but without pressure and without massage.

(*c*) Slight galls, chafes, or abrasions are treated with white lotion or powdered boric acid.

(*d*) Sitfasts are patches of dry dead skin and may involve deeper tissues. They are caused by continuous pressure of the saddle, cinch, or collar. When sitfasts appear, apply warm baths or warm poultices until the dead skin becomes loose. Then remove all dead and bruised tissue with forceps and a knife and treat with iodine and boric acid.

(*e*) If it is necessary to continue to ride an animal with a sore back, means must be devised to relieve pressure from the affected part.

q. Collar injuries.—(1) *Due to ordinary collar.*—Collar injuries are caused by friction which must be stopped if a sore shoulder is to be cured. One exception to this rule is the gall caused by pressure on the neck in front of the withers. Figure 12 shows the position of injuries produced by the collar. They are numbered from 1 to 5.

(*a*) Injury No. 1 is the gall just mentioned. It is an injury giving rise to extreme pain, and is more common in pole than in shaft draft. It is caused by the strain inflicted on the neck in pulling up or in going downhill, owing to the pull of the pole chains, aggravated by a loose housing strap. The injury may occur on perfectly level roads due to shortening the pole chains too much and so throwing the weight of the pole on the horse's neck.

(*b*) Injury No. 2 is due to a tight collar, especially in horses with fleshy necks. Remove some of the stuffing or stretch the collar and alter the hames accordingly.

(*c*) Injury No. 3 is due to a loose or wide collar. The remedy is to plug the collar (insert more straw) and so make it narrower. All movement in the collar should be stopped.

(*d*) Injury No. 4 is caused by a loose or a tight collar. The whole neck for about 5 inches is very tender, and the appearance of the collar is indicative of the cause. The remedy is to make it narrower with strips of numnah or wider by wetting

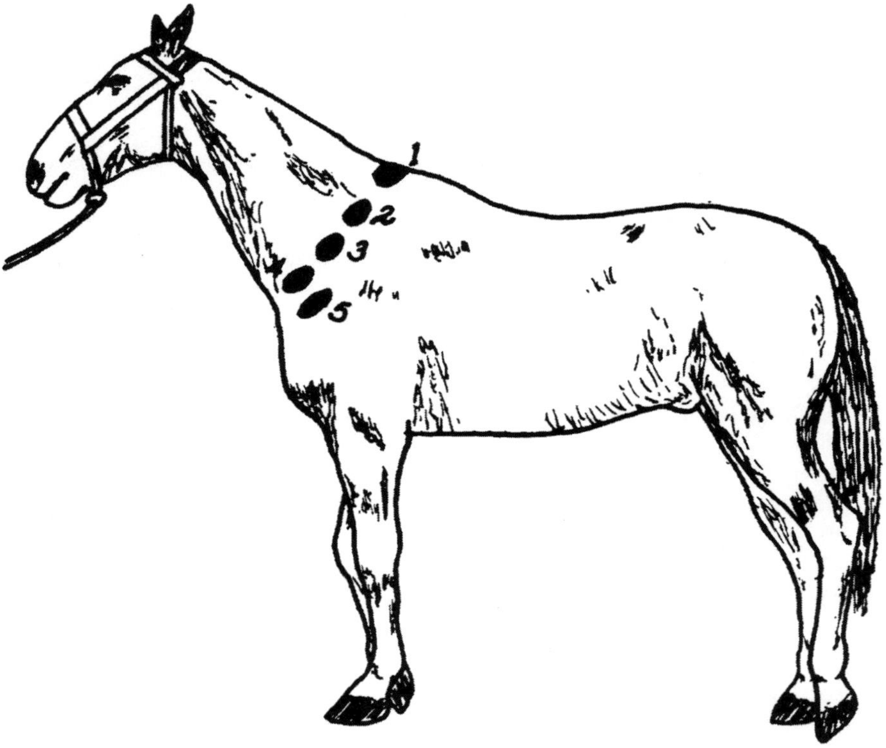

FIGURE 12.—Position of collar injuries.

it and stretching it on a block. It should be remembered that stretching a collar shortens it.

(*e*) Injury No. 5 occurs just above the shoulder joint and is due to the raw edge of the afterwale cutting into the skin. The remedy is either to restuff the collar or, if there is no time for this, to cut off the edge of the afterwale close up to the lacing.

(*f*) Sometimes an injury occurs in the region of No. 5 but above the edge of the afterwale. This is due to the friction of a wide collar, and the remedy is to deal with it as with No. 3.

(*g*) An animal with sore shoulders may be worked after the acute inflammation has subsided if the collar has been fitted.

(2) *Due to breast collars (field artillery harness)*.—(*a*) *Causes*.—Pressure or impact on the horse's neck caused by the downward pull on the wheeler's neck which results from a change in the direction of the line of draft at the point where the wheel and swing traces join; the weight of the pole being carried on the wheeler's necks; the impact on the wheeler's necks resulting from checking the carriage by means of the neck yoke strap and collar pad instead of the martingale, side straps, and breeching; poor driving or uneven gaits; dirty or rough collars (inside).

(*b*) *Prevention*.—Good draft is essential to the prevention of sore necks. There must be smoothness and steadiness particularly in starting the carriage, increasing and decreasing the gait, settling into hard pulls, and crossing obstacles. Prevent jerks and the consequent impact on the necks. The collar pad and the neck should be kept clean. The driver should discover any tenderness or sensitiveness of the neck, and steps should be taken to prevent a sore neck before it actually occurs.

1. *Hold-up straps*.—To relieve the wheel horse's neck from the downward pull of the traces and place it on the back, the traces are supported by hold-up straps fastened to the saddle and so adjusted that the traces in front of the straps are horizontal.

2. *Pole support*.—To relieve the wheel horses' necks of the weight of the pole, automatic pole support should be adjusted so that the pole is even with the bottom of the pulling muscle found at the top of the horse's arm, just in front of the elbow.

3. *Hippo straps*.—The hippo strap is a double or forked strap with a snap at the single end. When the snap is fastened in the pommel halter square, and the double or forked end is buckled to the collar body (in the same manner as the neck strap), the collar is supported in its proper position. By using hippo straps a horse with a very sore neck can be kept in draft and cured, since nothing touches his neck.

4. *Adjustment of breeching*.—Slackness in the breeching forces the wheel pair to hold back the carriage with their necks, through the neck yoke straps and collar pads. This pressure, particularly when received as a sudden blow, is one of the frequent causes of sore necks. A proper breeching adjustment involves correct placing and securing of the saddle, proper length of the back strap, correct height and slope of the breeching body, proper adjustment of the side straps and proper length of the martingale.

Section VI

SHOEING ANIMALS

152. General.—*a.* The purpose of horseshoeing is to protect the feet from breaking and wearing away without interfering with the normal functions of the feet.

b. The serviceability of riding, draft, and pack animals depends largely on the condition of the feet. The officer responsible for the serviceability of the animals should possess a knowledge of the construction and functions of the various parts of the foot. He should know when the feet are in need of shoeing, the shoes best suited for the work the animal must perform, the preparation of and balance of the feet, the fit of the shoes, and the securing of them to the hoof with the minimum amount of interference with the natural functions.

c. The horseshoer, in order to shoe correctly and intelligently, must have a working knowledge of the anatomy of the horse, and particularly the construction of the feet and legs.

d. A record of shoeing should be kept by each organization having animals. The actual period of time between shoeings should be governed by the needs of reshoeing in each individual case.

e. For detailed instructions concerning anatomy of the foot, duties of horseshoers, and corrective shoeing, see TM 2140–15.

153. Parts of the foot.—The horse's foot is composed of four parts:

a. The bones.

b. Certain elastic structures of cartilage and fibrous tissues (lateral cartilages and plantar cushion).

c. The corium, a layer of highly sensitive flesh or quick, which covers this bony and elastic framework.

d. The box, or case of horn, called the hoof, which incloses and protects the sensitive parts.

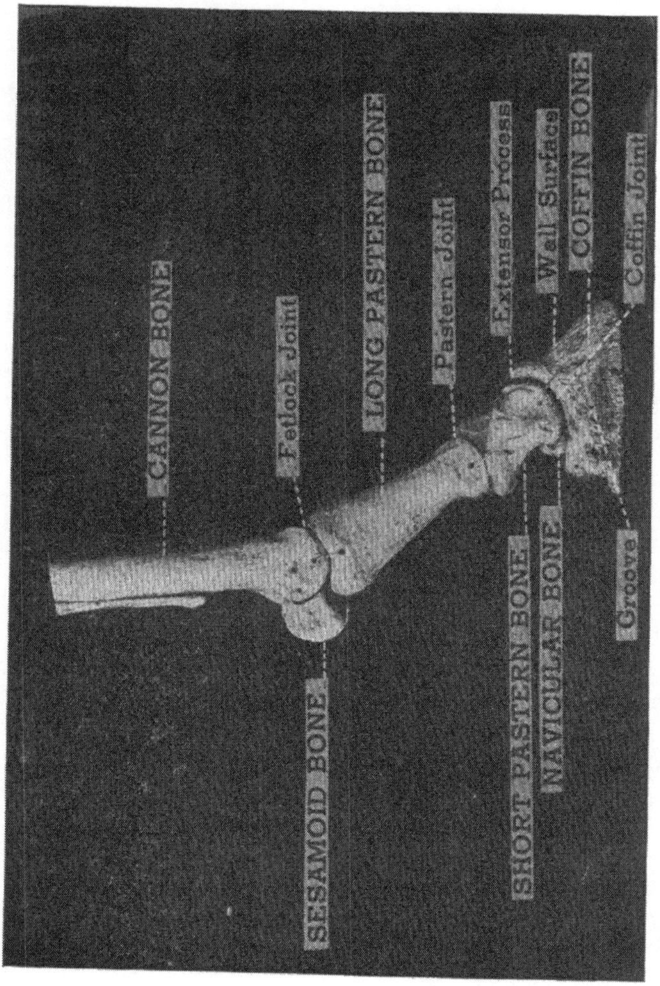

FIGURE 13.—Bones of the foot and pastern region.

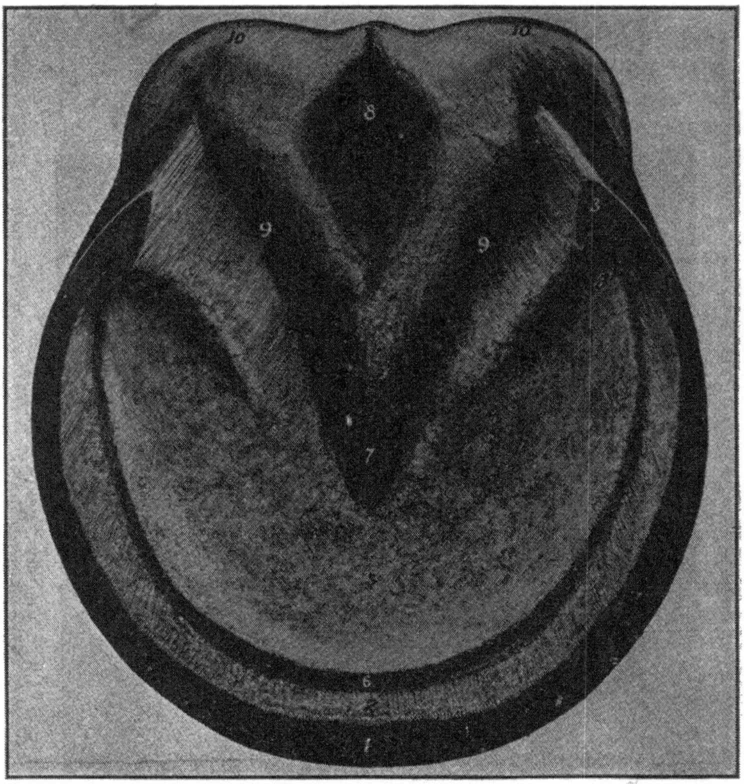

FIGURE 15.—Ground surface of foot.

1. Bearing surface of wall
2. Horny laminae
3. Angle of wall or buttress
4. Bar
5. Horny sole

5^1. Angle of sole

6. White line or junction of wall and sole
7. Point of frog
8. Cleft of frog
9. Commissures
10. Bulbs of frog

154. Frequency of shoeing.—*a*. The shoeing periods should be governed by the rapidity of the growth of the wall, the wear of the shoe, and the character of work done by the animals. Normally shoes remain on the feet from 3 to 7 weeks. The shoes should be removed and the surplus horn cut away whenever a growth of horn sufficient to affect adversely the natural functions of the various parts of the foot has accumulated.

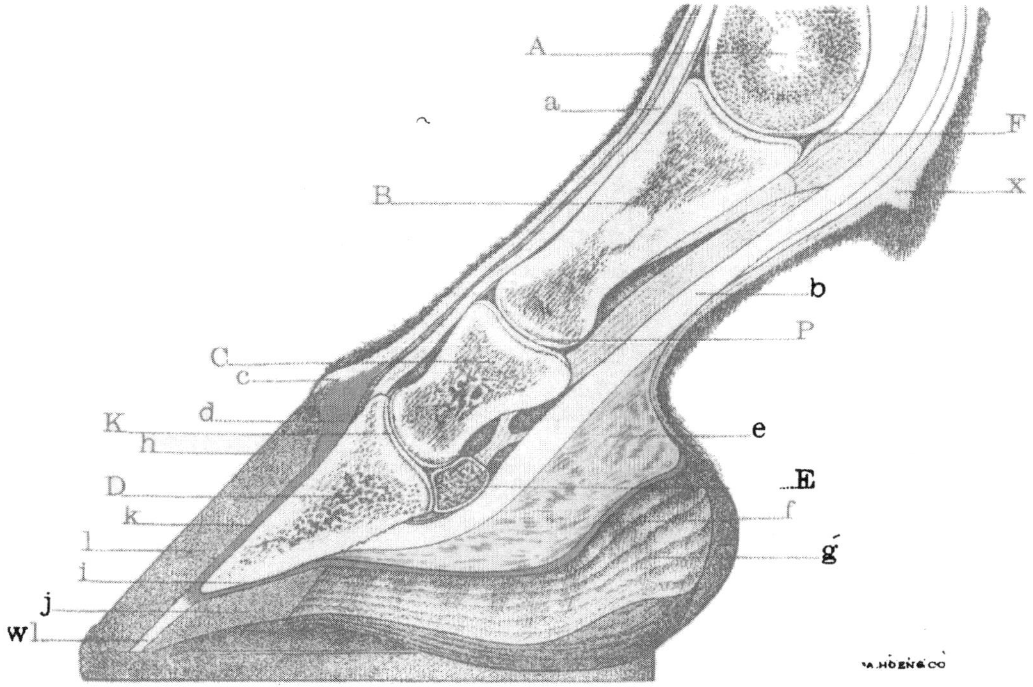

A. Cannon bone
B. Long pastern
C. Short pastern
D. Coffin bone
E. Navicular bone
F. Fetlock joint
K. Coffin joint
P. Pastern joint

a. Extensor tendon
b. Deep flexor tendon
c. Perioplic ring
d. Coronary band
e. Plantar cushion
f. Sensitive frog
g. Horny frog
h. Periople
i. Sensitive sole
j. Horny sole
k. Sensitive laminae
l. Horny wall
wl. White-line
x. Ergot

FIGURE 14.—Parts of the pastern and foot.

b. In some cases animals are shod too frequently thereby causing injury to the wall of the hoof by the increased number of nail holes in the wall. Riding horses used for garrison duty may go without hind shoes in the majority of cases. The unshod feet require trimming once a week to remove broken edges of horn and to level the foot.

c. (1) Suggestions for inspecting the feet and shoes for needs of reshoeing:

(*a*) Is the shoe worn out (unserviceable)?

(*b*) Are the clinches rough (projecting beyond the surface of the wall)?

(*c*) Is there any evidence of interfering (the inner side of the limbs cut or bruised)?

(*d*) Are the corresponding feet the same size, the toes the same length, and the heels the same height?

(*e*) Are the toes too long compared with the height of the heels (broken back)?

(*f*) Are the heels too high compared with the length of the toe (broken forward)?

(*g*) Is the wall of the hoof (from the coronet to the lower border) greater on the outside half than on the inside (broken in)?

(*h*) Is the wall of the hoof (from the coronet to the lower border) greater on the inside than on the outside (broken out)?

(*i*) Does the lower border of the wall extend over the shoe laterally (no provision for expansion in outline of the shoe)?

(2) The troop commander should personally inspect the feet of each animal in his command at least once weekly.

155. Normal shoes.—By a normal shoe is meant a shoe for a sound foot of a horse with proper gaits. The normal shoe may be used plain or with calks. The size of shoe required is determined by inspection, being careful not to take a shoe in which the last nail hole is in rear of the bend of the quarter of the hoof. Each size is made in five different weights: extra extra light, extra light, light, medium, and heavy. Shoes generally used are—

a. For riding and pack animals in garrison service, the extra light weight.

b. For riding and pack animals in field service, the light weight. The medium weight shoe in some cases will be most satisfactory for marching.

c. For draft animals in both garrison and field service, the medium weight.

156. Preliminary examination.—Before shoeing or reshoeing a horse, a careful preliminary examination of the gaits, the conformation of the legs, and the condition of the feet should be made in order to determine the correct method of shoeing to be used.

a. The animal should be examined to determine whether there is anything abnormal in his gait.

b. Place the animal on a level standing, view the feet and legs from a position in front of the animal and note the following:

(1) Are the bones of the pastern and foot in prolongation of the upper bones of the leg, or do they deflect inward (toe in)?

(2) Do they deflect outward (toe out)?

(3) Are the corresponding feet the same size in width of quarters and length of toe?

(4) Are the feet balanced laterally (the length of the wall from the hair line to the lower border equal on both sides of the foot)?

(5) Are the feet circular or narrow with pointed toes?

c. View the animal from the side and note the following:

(1) Are the feet balanced from toe to heel (the length of toe and height of heels in accordance with the conformation of the pastern)?

(2) Are the heels the same in height?

(3) Is the angle of the wall at the front part of the hoof (from the hair line to the lower border) parallel with the slope of the pastern?

d. Examine the wear of the old shoes as a check on accuracy of previous examination to determine balance of the foot and its position at the moment of breaking over and leaving the ground. If there is any malformation of the foot or evidence of irregularities in the gait, the type of shoe and fitting should be such as to correct the irregularity.

157. Preparation of foot and shoe for shoeing.—*a.* Cut the clinches of the nails and remove the old shoe, using care to avoid breaking away the wall. With the knife pare away the dead horn over the white line and particles of ragged horn on the frog. With nippers remove the surplus horn from the wall, then level the foot with a rasp. A covering of dead horn is essential to protect the sensitive sole against bruising. Some classes of work, such as marching over hard roads, require a greater amount of hoof than for ordinary work. To cut away the bars or open the heels weakens the foot and usually results in contraction of the quarters and heels.

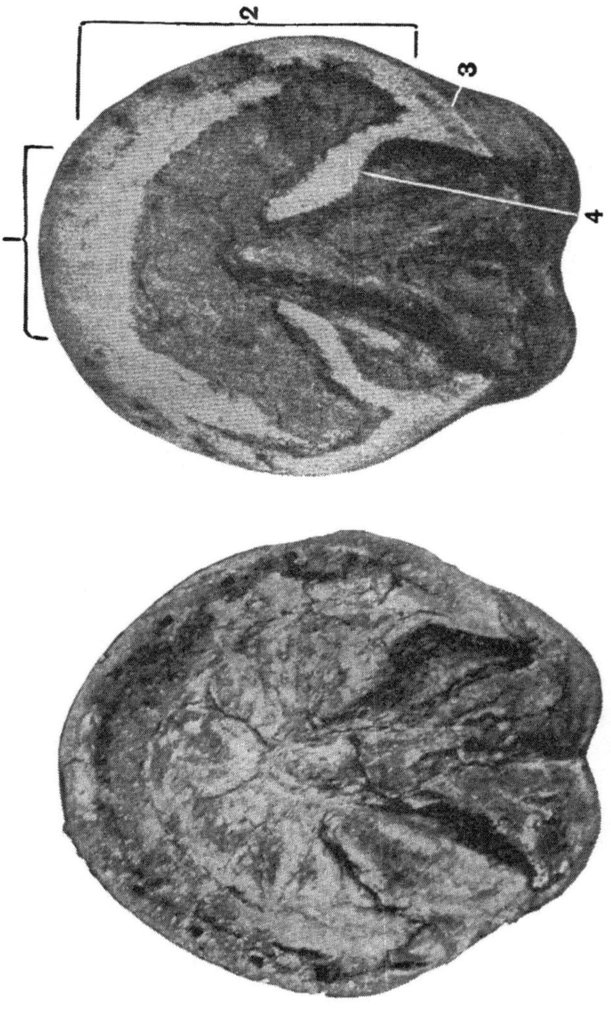

(1) Appearance of the foot after the old shoe has been removed.

(2) Foot ready for the new shoe. 1. Toe. 2. Quarters. 3. Buttress. 4. Bar.

FIGURE 16.—Preparation of the foot for shoeing.

b. The shoe should be fitted to afford the greatest protection to the hoof, minimum interference with the functioning of the foot structures, and to aid the feet to break over and leave the ground with correct position and rapidity to insure proper elevation and stride during flight. The shoe should be fitted with roundness (width) at the toe to give lateral support to the foot at the moment of breaking over and leaving the ground. A shoe pointed at the toe may cause the animal to interfere, stumble, or forge. The outline of the shoe at the quarters and heels should be sufficient to allow for expansion of the hoof and growth of horn between shoeing periods. The heels of the shoe should be of sufficient length and width to cover the buttresses, and finished smooth leaving no sharp edges. The upper and inner surface of the shoe should be concaved, as pressure of the shoe on the horny sole will cause lameness. The shoe should rest evenly on the bearing surface of the hoof.

158. To secure a fitted shoe.—*a.* The shoe is placed on the bearing surface of the foot and held firmly in position with the left hand, back up. A nail is placed between the thumb and forefinger of the left hand, the flat side of the nail faced outward. The nail is held squarely in the center of the most convenient nail hole, usually the third hole on the right side of the shoe. The nailhead is tapped lightly with the driving hammer. The nail should be held firmly to determine the feel, and the thumb and forefinger should not be removed until the nail needs no further guidance. The shoe is then grasped firmly and the nail is driven with slightly increased force. After the point of the nail emerges, turn it with the claw of the hammer until it stands at right angles to the wall. With the claw of the hammer engage the nail point and by quick half turn wring off the point close to the wall.

b. The second nail should be driven on the side opposite the first nail. When these two nails have been driven the foot should be allowed to rest on the floor, and the foot and shoe should be examined from all sides to see that the nails are properly placed and that the shoe is properly set. The shoer then drives the remaining nails, the fourth nails on either side being driven first. As each is driven it is promptly wrung off. When all nails have been driven the heads are forcibly hammered well into the crease.

c. The clinching block is now placed under one of the nail points and the nailhead is struck with the hammer. This bends the nail point over toward the bottom of the foot, and the nail is drawn up into the wall, thus tightening the shoe.

d. In seating the nailheads into the crease and in drawing up the nails and turning the clinches, the shoer begins on either side of the foot and takes the nails in the following order: the toe nail on one side, the heel nail on the same side, the toe nail on the opposite side, the heel nail on that side, and then the quarter nails.

e. After the clinches have been formed, the foot is brought forward upon the knees. With the rasp the clinches are made of equal length, being in length about the width of the nail at this point. The end of the nail is brought to a thin edge by filing off the corner of the underneath side. The clinches are then tapped lightly with the driving hammer until they are pressed against the hoof, then the outer edges are smoothed with the rasp.

159. Suggestions for inspection of a newly shod animal.—*a.* Are the corresponding feet the same size (toes same length, heels same height)? Has enough horn been removed? Has too much horn been removed?

b. Note the balance of the foot in relation to the limb, viewing from the front: Is the axis of the foot in prolongation of the axis of the upper leg bones? Viewing from the sides: Is the angle of the wall from the coronet to the toe parallel to the angle of the pastern?

c. Note the height and strength of nailing: Do the nails come out of the wall at a sufficient height and in sound horn to hold the shoe securely? Are the nails driven to a greater height in the wall than necessary? Was the proper size of nail used?

d. Note the strength and finish of the clinches: Are the clinches of sufficient thickness to insure strength? Are the clinches smooth and not projecting above the surface of the wall?

e. Note the size, weight, and type of the shoe: Is the quarter nail hole in the shoe at the bend of the quarter? Does the shoe conform in weight to the nature of work the animal performs and his ability to carry weight on the feet? Does the foot require a corrective type of horseshoe?

f. Note the outline of the shoe: Is the shoe fitted with sufficient fullness at the toe to give lateral support to the foot at the moment of breaking over? Are the branches of the shoe from the bend of the quarter to the heel fitted fuller than the outline of the wall to provide for expansion of the foot and normal growth of horn between shoeing periods? Do the heels of the shoe cover the buttress? Is the shoe finished smooth (no sharp edges)? Does the shoe rest evenly on the bearing surface of the hoof (wall, white-line, and buttress)? Is the shoe concaved (so that it does rest upon the horny sole)? Are the nailheads properly seated?

g. Note the rasping of the lower outer border of the wall (if much rasping is in evidence): Do the conformation of the foot and type of shoe used warrant the amount of rasping?

h. Note whether or not the ragged particles of the horny frog have been removed: Is there any evidence of improper use of the knife on the horny sole, horny frog, or bars? Is there sufficient frog pressure? If the toe clip is used, note whether it is in the center of the toe. Is it of sufficient strength, height, properly finished and seated?

i. Observe for lameness or faulty action of the horse at the various gaits.

160. Corrective shoeing.—*a*. Corrective shoeing is a term meaning the application of a shoe of special construction to correct for irregularities in gait or to correct and protect a hoof suffering from a foot disorder. The more common types of corrective shoes are bar shoe, bevel-edged shoe, calked shoe, rocker-toe shoe, and weighted shoe. For details of their construction see TM 2140–15.

b. The more common foot disorders which require corrective shoeing and the type of shoe recommended are—

(1) *Contracted quarters and heels.*—Bar shoe or bevel-edged shoe.

(2) *Corns.*—Bar shoe.

(3) *Thrush.*—Bar shoe.

c. The more common irregularities of gaits are stumbling, forging, and interfering. To correct these it is essential that the feet should be balanced. Rolled-toe shoes, rocker-toe shoes, and heel calks induce more rapid breaking over of the feet.

Square-toe shoes and lateral-extension-toe shoes induce the foot to break over at the center of the toe.

(1) *Stumbling.*—For stumbling in front, use the rocker-toe shoe. For stumbling behind, use a light-weight shoe with low heel calks.

(2) *Forging.*—Use rocker-toe front shoes and normal hind shoes.

(3) *Interfering.*—Use the interfering shoe described in paragraph 45, TM 2140–15.

d. The bar shoe used extensively in corrective shoeing is difficult to make with field equipment. A substitute for the bar shoe is easily made by turning the branches of the shoe inward as shown in figure 17.

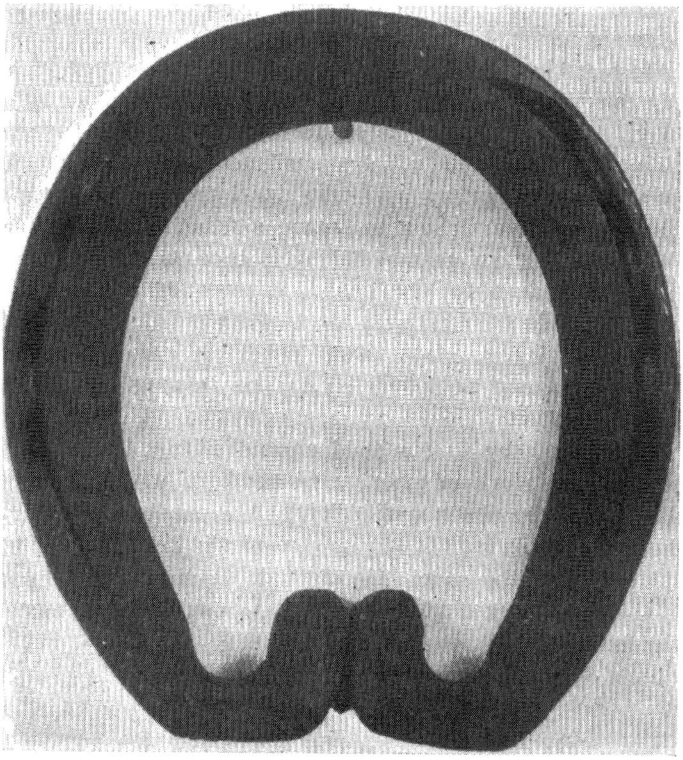

FIGURE 17.—Substitute for bar shoe.

161. Fitted shoes for road marches.—Organizations should maintain one front and one hind shoe for each animal, fitted and suitably marked, for use in the field. These extra fitted shoes should be kept in the shoeing shop, used at the next reshoeing of the animal, and replaced by newly fitted shoes.

162. Important field measures.—*a.* Troops should start into the field with animals freshly shod. One front and one hind fitted shoe with 16 nails for each animal are included in the equipment of the rider or driver. An extra supply of shoes and nails should be carried in the field train. Nails issued to individuals should be placed in a wooden container and oiled from time to time to prevent rust.

b. Any animal losing a shoe on the march must be given proper attention immediately. If neglected, lameness will almost surely result. Officers, noncommissioned officers, riders, and drivers must be constantly on the lookout for loose or cast shoes. The feet of animals will be inspected at every halt by the rider or driver.

c. A cast or loose shoe is secured to the foot as prescribed in paragraph 158. All officers and noncommissioned officers of mounted organizations and organizations equipped with animals must understand how to secure a fitted shoe. In case no shoe is available, an animal losing a shoe may be enabled to continue the march by the use of a piece of canvas, leather, or gunny sack as a covering to protect the foot.

d. As soon as possible after reaching camp the feet of each animal should be inspected by the responsible officer and by the horseshoer.

e. The horseshoer in the field should adjust his methods of performing the work of shoeing to conform with the facilities available. In the absence of a forge, a wood fire on the ground will heat the shoes sufficiently to cut off the heels, open the nail holes, and shape the shoe. He may, in the absence of an anvil, shape the shoe over a piece of rail or even over a wagon tire. Nailing and clinching in securing the shoe may be accomplished with tools designed for other purposes entirely.

Section VII

DUTIES OF ENLISTED SPECIALISTS

163. Stable sergeant.—The stable sergeant is responsible to the organization commander for the efficient care of the animals and stables of the organization (or picket line in the field) and all public property pertaining thereto. He should be quartered at the stables or in the vicinity of the picket line. In detail his duties are to—

a. Know the contents of this manual and of TM 2140-15.

b. Supervise and control the activities of all enlisted men on duty at the stables.

c. Take charge of the restraint and safe-keeping of all animals of the organization when they are not in use.

d. Take charge of the feeding and watering of the animals at the hours and in the amounts specified by the organization commander.

e. Execute the orders of the veterinarian in regard to the treatment of animals sick in stables and, in the absence of a veterinarian, to care for sick animals.

f. Receive, check, and care for the forage and bedding issued to the organization; issue the forage and bedding for feeding and bedding purposes and determine its fitness for use.

g. Take charge of the general police and minor repair of the stables and stable fittings.

h. Care for all transportation equipment, tools, etc., which are stored or habitually kept at the stables.

i. Keep the following records (par. 139):

(1) Descriptive card of public animals (or horse record card).

(2) Record of stable property.

(3) Forage record.

(4) Shoeing record.

(5) Morning report of animals.

(6) Sick report of animals.

j. Enforce the following rules for the care of animals at all times:

(1) Treat animals gently.

(2) Before entering the animal's stall and when coming up behind him, speak to him gently, then approach quietly.

(3) Never kick an animal, strike him about the head, or otherwise abuse him.

(4) Never punish an animal, except at the time he commits an offense, and then only in the proper manner, never in anger.

(5) Give an animal an opportunity to drink before leaving the stable or picket line and before putting the bit in his mouth.

(6) Never take a rapid gait until the animal has been warmed and circulation in the feet started by gentle exercise.

(7) When an animal is brought to the stable or picket line in a heated condition never allow him to stand uncovered. Put a blanket on him and rub his legs or walk him until he is cool. If he is wet, put him under shelter, not in a draft, and rub him with a wisp until dry.

(8) Never feed grain or fresh grass to an animal when heated. Hay will not hurt an animal however heated he may be.

(9) Never water an animal when heated, unless the exercise or march is to be immediately resumed. Sponging out the mouth and nostrils is refreshing to the heated animal and will not hurt him.

(10) Never allow an animal's back to be cooled suddenly. To cool the backs of riding animals gradually, remove the pack and equipment from the saddle, loosen the girth, remove the blanket, replace it with the dry side next to the animal, and fasten in place with the surcingle. With pack animals remove the load and later the pack saddle.

(11) Never put an animal up for the night until he is thoroughly cooled out and clean, especially around the legs, pasterns, and feet.

(12) Individual men returning from mounted duty or pass should report their return to the stable sergeant, who should inspect each animal and see that it is properly cared for.

(13) Never mount or ride an animal in a stable.

164. Enlisted assistants of stable sergeant.—The stable sergeant's assistants are the horseshoers, the stable orderly or orderlies, and such other men as may be detailed for special duty at the stables. The horseshoers and the stable orderlies should, when practicable, be quartered at the stables or in the vicinity of the picket line in the field.

a. Horseshoers.—In detail, the horseshoers' duties are to—

(1) Know the contents of this manual (especially those parts pertaining to shoeing animals) and of TM 2140–15.

(2) Shoe the animals designated by the stable sergeant.

(3) Inspect the feet of animals daily, preferably at "Stables," and attend to any that need it.

(4) Maintain the shoeing shop in a good state of police.

(5) Care for the tools issued for horseshoeing purposes.

b. Stable orderlies.—The stable orderly or orderlies and other men detailed for special duty at the stables police the stables, make minor repairs to the stables, especially the stall floors, and perform such other duties connected with the animals or stables as the stable sergeant may direct. At least one man must be kept on duty at the stables day and night.

c. Veterinary Corps personnel.—Enlisted men of the Veterinary Corps who may be attached to the organization are under the control of the stable sergeant for administrative purposes, but carry out the instructions of the veterinarian in regard to the care of sick and injured animals of the organization.

SECTION VIII

TRANSPORTATION OF ANIMALS

165. Rail.—*a. Type trains, mixed trains, or freight trains.*—The use of type trains, adapted in passenger and cargo space to the organizations they are intended to carry, facilitates the movements of troops in or near the theater of operations. With the long hauls of the zone of the interior, the usual fast schedules cannot be maintained if the trunk lines are encumbered with the slower-moving type trains. For this reason it may be necessary to limit the use of type trains to those situations in which it is expected that the troops will be employed immediately upon detraining. The initial concentrations and most movements entirely within the zone of the interior will be executed by using separate passenger and freight trains often moving over different roads. These can be assembled at or near their destination so as to permit men, equipment, and animals to detrain at the same time. It will usually be necessary for freight and animals to be sent on in advance. Where an organization is too small to warrant complete passenger and freight trains, a mixed train may be made up but, as such trains are slower moving than the usual passenger traffic, their use is attended with the same disadvantage as that of the type train. Their use should, therefore, be kept to a minimum.

b. Maximum cars in single shipment.—A single shipment should not exceed 30 cars, as there are few rest stations where more than 30 carloads of animals can be unloaded at one time.

c. Halters and tie ropes.—Halters and tie ropes should ordinarily be carried in sacks by the attendants, but halters may remain on the animals, the tie ropes being carried in sacks.

d. Forage.—(1) *Less than carload.*—In less-than-carload shipments of animals, whenever there is room for forage in the car, the necessary forage for the entire journey will be furnished by the Quartermaster Corps and loaded in the car at the initial point of shipment.

(2) *One or more carloads.*—In shipments of one or more carloads of animals, or in less-than-carload shipments whenever there is no room for forage in the car, forage will not be furnished by the Quartermaster Corps but will be supplied by the carriers at their regular feeding points, except when the animals are accompanying troops, in which case the necessary forage will be carried in boxcars attached to the train.

e. Preparation of cars.—The health of the animals is the primary consideration, and shipment should be made only in cars that have been properly cleaned, disinfected, and sanded.

(1) *Cleaning and disinfection.*—(*a*) *By whom done.*—The necessary and usual cleaning and disinfection of cars must be done by the carrier without expense to the Government. Any additional disinfection that may be ordered will be done at Government expense.

(*b*) *How done.*—When disinfected by other than the carrier, the disinfection ordinarily will be done by the Veterinary Corps, or under their direction. The car is sprayed with cresolis solution prepared by adding one-half mess-kit cupful of cresolis to one galvanized iron bucket nearly brimful of water.

(2) *Sanding.*—Sanding cars will be done at Government expense.

(3) *Papering.*—For the better protection of animals in severe weather one side of ordinary stock cars, and also both ends,

if not solid, should be closed with heavy paper or slats fastened on the outside of the cars so as to avoid projecting nails if torn off. Burlap from sacks or salvage canvas may be used in lieu of paper. It will be done at Government expense.

(4) *Advance charges.*—The cost of disinfecting, sanding, or papering done at Government expense will be treated as an advance charge against the bill of lading.

(5) *Inspections.*—(*a*) *By a veterinary officer.*—Stock cars will be inspected on arrival and also when ready for loading by a veterinary officer, who will advise the consignor as to any sanitary defects to be corrected and will supervise any additional disinfection that may be ordered.

(*b*) *By the consignor.*—Prior to loading, all stock cars will be thoroughly inspected by the consignor.

1. Hayracks and water troughs must be in good repair.
2. Special attention should be given to the condition and repair of the floor.
3. Projecting nails must be removed; they can best be detected by looking along the surface instead of at it from the front.

f. Loading.—(1) *Number of animals to a car.*—(*a*) *Thirty-six-foot car.*—The following are the maximum numbers of animals to be placed in a 36-foot car:

Heavy draft horses	18
Light draft horses	20
Riding horses	22
Draft mules	21
Pack mules	22

In very hot weather these numbers may be reduced by two. Weather conditions permitting, animals will ride better and with less injury in cars if they are sufficiently closely loaded so that each animal receives some lateral support from the bodies of adjacent animals.

(*b*) *Forty-foot car.*—When 40-foot cars are used, the number of riding horses and pack mules given in (*a*) above may be increased by three, and the number of heavy draft horses, light draft horses, and draft mules may be increased by two. The majority of livestock cars now in service are 40-foot cars.

(2) *Methods.*—(*a*) Animals may be most conveniently loaded through a stockyard chute or from a freight platform level with the floor of the car; when these are not available, portable or improvised ramps must be used. When a ramp is used it must be well supported and have strong sides.

(*b*) The loading should proceed without noise or confusion, the animals being led quietly to the car door and there turned over to four men, two for each end, who do the loading. They should follow one another promptly in order to avoid delay, and should be kept moving so as to reduce the tendency to balk. They should be led in single file by means of halter straps or tie ropes, which should not be removed until the animals are in place in the car. In some cases it may be necessary to blindfold an animal before he can be led into the car. A rope or strap passed in rear of the haunches and drawn forward by a man at each end is often effective in urging an animal into the car. The first animal loaded should be led to one end of the car, the second animal to the other end, and so on, leaving the center of the car for the last animals loaded. Except as prescribed in (*c*) below, arrange the animals so that alternate ones face in the same direction. This facilitates loading by conserving space rather than with any idea that animals will long maintain this position, for once the car is under motion, the animals will move about considerably. The gentler animals should be placed opposite the doors and should, therefore, be loaded last. Before starting to load, the door on the farther side of the car must be closed and fastened. As soon as the loading is complete the opposite door must also be closed and fastened. Adequate ventilation must, however, be provided for at all times.

(*c*) Teams of mules should be loaded and should stand in the cars as they are driven together in the team. Mules should be tagged with numbers so as to be quickly identified. Horses accustomed to service together should be loaded together and in the same car. If necessary for identification they should be tagged.

(*d*) Sick or injured animals should be loaded in a car by themselves and a separate stall built for each animal, if practicable, unless palace stock cars are furnished.

g. Rest and feeding en route.—Shipments will be so handled that animals will not be kept in cars for a period longer than 28 consecutive hours without being unloaded (act June 29, 1906 (34 Stat. 607)), except in extreme cases of unavoidable delays or military necessity.

(1) *Release; when signed.*—The Government representative in charge of a shipment will sign a release only in these extreme cases.

(2) *Time limit.*—Animals should be unloaded, fed, watered, and rested after 20 hours, or even less, rather than risk exceeding the 28-hour limit. In no case will more than 36 consecutive hours be permitted to elapse without unloading for rest and feeding.

(3) *Period of rest.*—Animals will be allowed not less than 5 consecutive hours of rest at each unloading point.

(4) *Notice to railroad authorities.*—When disinfection of feed and rest station facilities is desired, a written request for such service should be supplied the carrier's agent at the point of origin of the shipment. Carriers are expected to maintain such facilities in a clean condition, but requested disinfection is done by the carrier at the shipper's expense.

(5) *Notice to Bureau of Animal Industry, United States Department of Agriculture.*—A copy of the request referred to in (4) above should be mailed to the appropriate bureau inspectors in charge, if such are available, requesting that if possible that agency supervise the disinfection done by the railroad company.

(6) *Inspection and cleaning of mangers and troughs.*—Upon arrival at a feeding station and before unloading the animals, the attendant will make a careful inspection of all mangers and water troughs. The proper cleaning of such mangers and troughs is done by the railroad company without expense to the Government.

(7) *Watering.*—Water troughs should be drained before the animals are unloaded, and they should not be allowed free access to water for at least 2 hours after being fed hay.

(8) *Feeding.*—Enough hay will be purchased at each feeding point to keep hay constantly before the animals while they are in the corrals. Hay will not be placed in cars or animals fed in cars in transit unless the animals are in individual stalls.

(9) *Guard.*—All forage, mangers, and water troughs should be kept under close supervision at the time of feeding, and guard will be kept at all times while the animals are in the corrals.

(10) *When unloading not necessary.*—When the stock cars provided are such that the animals may be fed and watered on the train, it is unnecessary to unload them for exercise or recuperation unless the weather is very hot and the journey long.

h. Unloading.—The door should not be opened until the car is opposite the unloading point. Each animal should be led to the door so that his body will be athwart the car before he starts to leave it.

166. Water.—*a. Kind of vessel required.*—For the transportation of animals by sea the transport service should be equipped with large steamers provided with bilge keels. The interior fittings should be of the most substantial character, the construction and arrangement of the stalls being in accordance with approved plans on file in the office of The Quartermaster General.

b. Instructions and specifications for conversion of commercial vessels.—Specifications for hasty conversion of commercial vessels for use as animal transports and forms for contracts are on file in the office of The Quartermaster General; in the office of each general superintendent, Army Transport Service; assistant general superintendent, Army Transport Service; or local quartermasters near important ports in the United States. Additional copies may be furnished by air mail where needed. As these specifications are based on the use of standard commercial fittings and materials readily secured locally in shipping centers, the acquisition and conversion of vessels may be expeditiously carried out either by the corps area commander or by telegraphic instructions from the War Department.

c. Equipment required.—Transports carrying animals will be equipped with the following for each 100 animals:

6 stable brooms	4 manure baskets
6 stable forks	4 hoes
6 buckets	4 brooms
4 shovels, short handle	200 hay nets

d. Inspections before sailing.—The port remount officer accompanied by the port veterinarian will inspect animal transports as follows:

(1) *Upon arrival.*—He will inspect every animal transport upon its arrival in port, after which inspection he will recommend to the port commander any changes or alterations to be made and the number of animals to be loaded. During his inspection he will pay special attention to the following points:

(*a*) The ramps and other facilities for loading.

(*b*) The accommodations for the animals, including the type and suitability of the fittings, adequacy of head room, space for each animal, space for cleaning, feeding, removal of sick or injured, method of tying in stalls, supply of hay nets and feed boxes, provisions for insuring and supplying an adequate supply of fresh air, both by artificial means and auxiliary ventilating apparatus such as wind sails and suction or blower fans.

(*c*) The arrangement for cleaning and flushing all decks.

(*d*) Lighting, including lanterns and candles, in case of failure of the electric lighting system.

(*e*) The water supply, adequacy and provisions for renewal, and the cleanliness of storage tanks; the forage, as to quantity, quality, and storage facilities.

(*f*) The supply of thermometers for each hold in which animals are carried; the supply of cleaning and grooming equipment.

(2) *Day before sailing.*—On the date preceding sailing of the vessel he will make another inspection and submit a written report to the port commander covering the following points:

(*a*) Number of animals to be shipped.

(*b*) Have specifications for fitting, etc., been carried out in a satisfactory manner?

(*c*) Are there 5 percent sick stalls on each deck?

(*d*) Is ventilation of horse decks satisfactory?

(*e*) Is drainage of horse decks satisfactory?

(*f*) Is water service satisfactory?

(*g*) Is lighting service satisfactory?

(*h*) Is ventilating apparatus satisfactory?

(*i*) Amount of forage aboard.

(*j*) Is storage of forage satisfactory?

(*k*) Is supply of cleaning equipment satisfactory?

(*l*) Is supply of grooming equipment satisfactory?

(*m*) Is supply of veterinary medicines, etc., reported as satisfactory by the transport veterinarian?

(*n*) Number of attendants, both remount service and veterinary service.

(*o*) Are precautions against fire satisfactory?

(3) *Immediately before sailing.*—After animals have been loaded and immediately before sailing of the vessel he will make a final inspection and will submit a written report to the port commander covering the following items:

(*a*) Name of transport remount officer in charge of animals.

(*b*) Name of transport veterinarian.

(*c*) Number of animals loaded, by classes.

(*d*) Is the ship in all respects ready for sea?

e. Transport remount officer.—(1) On each animal transport there will be assigned one officer of the remount service as transport remount officer; one senior noncommissioned officer, Quartermaster Corps, as first sergeant; one noncommissioned officer, Quartermaster Corps, as forage master; one cook for each 20 attendants; one noncommissioned officer and 13 privates of the remount service for each 100 animals; one veterinary officer as transport veterinarian; and enlisted men, Medical Department (Veterinary Corps), as follows: one staff sergeant, one private or private, first class, for each 100 animals, and one additional noncommissioned officer to each four privates or privates, first class. This personnel will, as far as possible, be permanent as long as the transport to which they are assigned is employed for the shipment of animals. When the number of animals exceeds 1,200, an additional veterinary officer as assistant to the transport veterinarian may be assigned.

(2) Under direction of the port remount officer the transport remount officer will have direct charge of the loading of the animals on his transport. He will have charge of all animal attendants on board and will assign officers and men to appropriate duties. He will have direct charge of all animals on board. When additional personnel is necessary for care of the animals he will make request for same to the commanding

officer of troops aboard the transport. Upon return from each trip he will make written report to the port commander, through the port remount officer, covering the following:

(*a*) Name of transport.

(*b*) Name of transport veterinarian.

(*c*) Names of attendants.

(*d*) Number of animals, by classes, loaded.

(*e*) Number of animals, by classes, delivered at destination.

(*f*) Number of animals, by classes, died, giving reasons.

(*g*) Condition of animals upon unloading.

(*h*) Nature of voyage.

(*i*) Quantity and quality of forage furnished.

(*j*) Water supply.

(*k*) Ventilation.

(*l*) Fitting of ship; if not satisfactory, detailed recommendations to be submitted.

(*m*) Equipment; if not satisfactory and complete, detailed recommendations to be submitted.

f. Transport veterinarian.—(1) The transport veterinarian will have charge of all veterinary attendants and will assign officers and men of the Veterinary Corps to appropriate duties. He will be constantly present during the loading of the transport. He will report promptly to the port veterinarian any defects in sanitary conditions which should be corrected before loading. He will, under the transport remount officer, have charge of the sanitation of that part of the ship occupied by animals. He will act as veterinary adviser to the transport remount officer. He will be responsible for the proper care of all sick or injured animals and will request their removal to such places as he may consider advisable. He will be responsible for the proper economical use of all veterinary supplies and equipment. On arrival at the home port he will take prompt steps to renew supplies and repair or replace equipment, to the end that both may be adequate and serviceable at all times.

(2) He will familiarize himself with quarantine regulations of foreign and home ports and with the prevalent communicable animal diseases thereat in order that he may advise necessary measures to prevent the admission and limit the extension of such diseases.

g. Attendants.—(1) The senior noncommissioned officer, Quartermaster Corps, exercises a general supervision and acts as first sergeant of the Quartermaster Corps detachment on board.

(2) Attendants, at the rate of not less than one for each 50 animals, will be constantly on duty with the animals day and night.

h. Animals not accompanying organizations.—In general, unless an organization accompanies a shipment of its own animals, it will turn into the remount depot at its station all of its animals and will draw new animals from the remount service after arrival overseas. However, if it is desired to reissue to any organization its own animals after arrival overseas, and the organization does not accompany the animals, the supply company or a detachment of the organization, with a sufficient number of men attached to furnish one man to every 10 animals, will be left to care for animals in the stables or animal shelters at the station of the organization until such time as the animals are sent to an embarkation depot. Where there is a remount depot the facilities thereat will be used to full capacity for caring for these animals awaiting shipment. The supply company or detachment with attached personnel will accompany and care for the animals while en route to embarkation depot, while at the embarkation remount depot, and while on board ship en route overseas.

i. Animals accompanying organization.—(1) *Shipped in advance.*—When animals of an organization can be accommodated at embarkation remount depots in advance of the date of departure of the organization for a port of embarkation, and it is intended that the animals accompany the organization, the animals will be shipped at least 2 weeks in advance of the date of departure of the organization, in charge of the personnel prescribed in *h* above.

(2) *Loading and care while on transport.*—When organizations proceed on the same animal transports with their own animals, the organization commander will, under the port remount officer, take charge of the loading, and the animals will be cared for while on the transport by the personnel of the organization.

j. Preparation for shipment.—As far as military requirements will permit, no animal will be shipped without having been held at least 2 weeks in an embarkation remount depot. The shipment to an embarkation remount depot for transportation overseas of sick or unserviceable animals is prohibited. When ordered to send animals to a port of embarkation, the commanding officer of the organization to which the animals are issued will submit all sick and unserviceable animals to the action of an inspector, who will recommend that such animals as he finds unfit for shipment to a port of embarkation, but not to be destroyed, be turned in to the nearest remount depot to be sold or continued in service, and those which should be destroyed will be disposed of by the responsible officer in accordance with the provisions of Army Regulations. The animals so condemned or ordered turned in will be immediately replaced by the issue of healthy and serviceable animals from the reserve supply at the remount depot. Before shipment the manes of all public animals will be roached closely, and animals to be shipped to the Tropics should be clipped. Animals will be freshly shod in front only, with smooth shoes, unless intended for immediate use after debarkation, when they may be shod smooth all around. Before loading, animals will be given soft feed for at least 2 or 3 days, and for 24 hours previous to shipment the forage rations will be cut in half. Animals will be watered immediately before loading, unless loading is to be by slinging, in which case no water should be given for 4 or 5 hours previous to loading.

k. Loading.—(1) The additional personnel required for loading will be furnished by the port remount officer from the personnel on duty at the remount depot. With the transport at the wharf, the animals are led on board by ramps or they are hoisted by means of slings or flying stalls. The ramps, decks, etc., should be covered with cinders, sawdust, or litter.

(2) In loading by ramps the animals should be led quietly, without interruption, starting with a gentle animal. Those that offer resistance should be set aside and later, if necessary, forced up the ramp by a rope passed in rear of haunches, or by tying a loop of a long rope around the animal's body over the loins and carrying the end under the belly, between the forelegs, through the halter, and thence to two or three

men some distance up the ramp who exert the necessary traction on the rope. On reaching the stable deck the animals will at once be led to the farthest vacant stalls where a feed of hay should be ready for them in a hay net.

(3) When animals are slung, all the apparatus will be carefully inspected and great care will be taken to prevent injury in hoisting or loading. When all is ready and "Hoist away" is given, the animals should be hoisted steadily and rapidly to the required height and then carefully lowered.

l. Debarkation.—(1) Additional personnel required for unloading will be furnished from the remount depot at the port of debarkation. Unloading will be in charge of the transport remount officer.

(2) With the transport alongside a wharf, the animals may be led off the ship by use of ramps or hoisted and lowered by means of slings or the flying stall. Unloading by the use of long ramps is always preferable to the use of flying stalls or slings as the work can be done much faster and more safely.

(3) A bed of sawdust, soft earth, or straw should be prepared to prevent injury to the knees of animals which may fall upon landing. Animals will be received by men on shore, and when a sufficient number has landed they will be led to the stables, picket lines, or corrals provided.

(4) When it is necessary to lower horses into lighters the greatest care must be taken, especially if there is a swell, to prevent injury. The horse should be received carefully on a bed of straw, by several men, and the tackle must be slackened down rapidly or let go altogether as soon as he strikes the boat in order that he may gain his footing.

(5) In smooth water it is best to construct a ramp from the ship to the lighter. This will facilitate unloading and save risk of injury.

(6) Animals may be landed by swimming, being lowered into the water by means of a sling or the flying stall, or they may be led to a port and forced into the water. The latter method, when practicable, is safer and quicker. A gangplank constructed of boards 16 feet long, the crosspieces being nailed on the under sides, is poised on the edge of the port. The animal is led to the port, a light rope is tossed up from a boat below and passed through the halter ring; the animal, after being

led onto the gangplank, is forced into the water by lifting up the interior end of the plank. When he begins to swim he is pulled up close to the boat, care being taken not to hold his head too high for swimming. At the beach the rope is given to a man who wades out from shore and takes charge of the animal.

m. Routine on transport.—The ship will be divided into sections, to each of which a noncommissioned officer with a squad will be assigned. Stalls will be numbered and the limits of the sections accurately defined so that each noncommissioned officer will know exactly for what animals and space he is responsible. The noncommissioned officer in charge of the squad makes his own details, keeps a forage record, and notes on the bulletin board the names of the men on guard and the number of horses sick.

(1) *Ventilation.*—(*a*) Too much importance cannot be attached to the provision of ample means for ventilation. It is at all times difficult to ventilate lower decks and holds which are beneath the water line, and special care should be taken to provide methods both for withdrawing the foul air and for pumping in fresh air; this is especially necessary in the vicinity of stalls which are farthest removed from the neighborhood of the hatchways.

(*b*) In addition to hatchways and portholes, which should be kept constantly open whenever possible, the means for ventilation are—

1. Permanent air funnels.
2. Iron wind scoops.
3. Canvas wind sails.
4. Electric fans or blowing machines.

(2) *Feeding.*—(*a*) No hard and fast rules as regards feeding on board ship can be prescribed. The following will serve as a guide:

1. Feed no oats for first 2 days.
2. For the next 2 days feed about 2 pounds of oats and 1 pound of bran per day to each animal.
3. Thereafter increase to, but do not exceed, 3 pounds of oats daily and 1 pound of bran.

4. Bran mashes and hay may be given freely throughout the voyage.

5. Not to exceed ⅓ pound of linseed meal should be fed each animal daily. This may be fed dry or as a mash.

(*b*) The greater the heat that is encountered the more should the oats be cut down and a greater reliance placed on hay and bran. The consequences of overfeeding with oats on board ship are disastrous, as digestive disarrangements are certain to follow.

(3) *Watering.*—(*a*) A supply of pure drinking water of not less than 10 gallons per day for each animal must be provided. Animals will be watered not less than four times a day; before being fed in the morning, at noon, before feeding in the afternoon, and again about 8 o'clock in the evening.

(*b*) If water is freely given and no horse is ever allowed to suffer for a moment with thirst, the good condition of horses on the ship is assured. If given an abundant and frequent supply of water the animals will maintain and even put on flesh aboard ship on a diet of hay alone. It is generally safe to assume that a consignment of animals arriving at a port of debarkation in markedly poor condition has been neglected as regards watering. If the feed boxes are watertight it is a good plan, after the final watering in the evening, to leave the boxes full of water, to be taken down during the night, when empty, by the guard.

(4) *Grooming and exercise.*—These will be carried out wherever and whenever possible. Every animal will be taken out of his stall at least a few minutes each day and moved about before being groomed. Hand rubbing the legs will help to offset the lack of exercise.

(5) *Bathing.*—Use hose pipe with sea water twice daily, when possible, for all animals suffering from heat. This lessens sweating, conserves condition, and purifies the atmosphere.

(6) *Sick animals.*—(*a*) *Space for.*—Not less than 5 percent of the stalls on each deck near the hatchways will be reserved for sick animals. As forage is used, other space will become available for ailing animals.

(*b*) *Treatment.*—The sooner a sick animal is detected and removed to an airy upper deck the better will be the chances of recovery and the less the spread of infection. Ordinarily no attempt should be made to treat hopeless cases, as it is contrary to hygienic principles and may lead to infection of neighboring animals.

(7) *Foothold.*—To insure a satisfactory foothold, ashes will be sprinkled freely on platforms in rough weather and along alleyways whenever an animal must be moved.

(8) *Policing.*—(*a*) *Outgoing voyage.*—The ship will be thoroughly policed each morning. All manure will be removed and decks flushed down. Each stall will be cleaned while the horse is being exercised and groomed. Feed boxes, stalls, and passageways will be washed out each day. Disinfectants will be used as required. Where ammonia gas is bad on account of poor ventilation, pieces of sacking saturated with water and sprinkled with a solution of chloride of lime will be hung up in places where the ammonia is most evident and kept spread out so as to present the largest possible surface to the air.

(*b*) *Return voyage.*—When transports return empty they will be thoroughly cleaned and disinfected during the return voyage so that loading may not be delayed after arrival. All of the attendants will be considered available for this duty.

(9) *Daily inspection.*—That portion of the ship occupied by animals will be thoroughly inspected at least once a day by the transport remount officer and transport veterinarian. The stalls and passageways must be cleaned and the scuppers clear. The noncommissioned officers in charge of squads will be present when that portion of the ship assigned to them is inspected.

167. Motor.—*a. Vehicles available.*—Military vehicles best adapted to employment in transporting animals are the four-wheel trailers or semitrailers which have a low center of gravity. These vehicles are not subject to undesirable side-sway when loaded with animals as are some of the standard cargo trucks. Loading and unloading of animals are greatly facilitated when using a vehicle of the low-slung body type. However, any of the standard cargo trucks may be used provided they have sufficient weight-carrying capacity and bodies of sufficient size to accommodate one or more animals.

134205°—39——11

b. Necessary modifications of vehicles.—Most of the standard cargo type bodies when used for animal transport must be modified by the attachment of a specially built rack or stake body which will prevent animals from jumping or falling out of the vehicle. This stake body should measure at least 4½ feet in height from the floor of the truck. The framework must be sufficiently strong to withstand the weight of a full load of animals surging against the sides of the vehicles when operating around curves or on turns. It should be so constructed as to permit ample ventilation. Since the rack or special body must be firmly fixed to the standard body of the truck, this usually necessitates bolting or welding to the framework of the regular body. All sharp edges or projecting surfaces of the inside of the body must be padded to prevent injury to the animals. This may be done by tacking hay-stuffed canvas pads over the surfaces which might cause injury.

c. Loading ramps.—Animals are loaded or unloaded from the rear of the vehicle. The rear end of the body of the vehicle may be so constructed that it will open out as a ramp for loading and unloading, in which case it should be so built that its slope when in use is not greater than 25° from the horizontal. The ramp should be the full width of the rear of the vehicle. If constructed of wood, it should be cleated with horizontal strips attached 6 inches apart over the entire surface of the ramp. If the ramp is made of steel, it should be cleated with permanently affixed steel strips similar to the ones used with the wooden ramp, or entirely covered with canvas-rubber fabric to prevent slipping of animals when being loaded or unloaded. In posts or at points where animals are frequently loaded or unloaded for motor transport, a permanent earth or masonry ramp is usually constructed. These ramps should be so located and designed that animals can be easily loaded in vehicles backed up to the ramp, or unloaded the same way.

d. Methods of loading.—(1) The manner in which animals are placed in a motor vehicle for transport will depend on the size and shape of the vehicle used and the number of animals to be loaded. Some vehicles permit loading with the animals standing side by side, all headed in the direction of travel. This is the most satisfactory method and is always used where the shape and size of the body of the vehicle permit the economical utilization of space available. When this method is used, either one or, in the case of large semitrailers, two rows of animals are loaded. A stout bar of 2 inch by 8 inch material is placed horizontally across the body of the vehicle in front of the breasts of the animals. Halter shanks are tied to these bars to prevent animals from rearing or lying down.

(2) Tight loading and economical utilization of all space available in the vehicle are important. Animals should be closely loaded so as to support each other when the vehicle sways, thus preventing falls and consequent injuries. If the number of animals to be transported does not fill the body of the vehicle, they should be packed tightly in one end and a bar placed across the body of the truck to restrict movement of the animals.

(3) Some vehicles are of such shape and size that it is best to load animals crosswise, packed tightly from front to rear. The old style "Class B" truck is an example of this type of vehicle. In this case animals are loaded "head and tail"; first animals in the truck headed to the right, next to the left, and so on. Heads of animals may project over the side, in which case halter shanks are tied on the outside of the body of the truck and sufficiently low to prevent animals from rearing.

(4) In cases of vehicles frequently used for transportation of valuable animals, small individual stalls are sometimes erected in the body of the truck. This is the safest method of all but, if used, stalls should be sufficiently narrow to support animals when sidesway occurs, otherwise they may become "cast" in the stalls and seriously injured.

(5) Floors of all vehicles used in animal transport should be sanded prior to use. In the case of long movements this sanding is repeated, when necessary, at intervals during the journey. Vehicles with steel floors are particularly conducive to slipping and falling of animals during transport. In cool weather these floors are sometimes covered with a thin coating of asphalt which is sprinkled while warm with sand. Some floors have curved wheel wells projecting from their upper surfaces. Animals should be so placed in these vehicles that

they are not standing directly on these wells, and wells always should be covered with some material like pieces of old innertube or canvas which will prevent the animal from slipping and struggling every time the wheel wells are stepped on.

(6) Loading and unloading animals on motor vehicles should proceed quietly and without confusion. Timid animals must be handled with patience. In loading, a man leads the animal up the ramp using the halter shank; a second man follows the animal and encourages him to go forward by "clucking." In cases of difficult animals, the man leading is assisted by two men in rear who, by pulling on the ends of a halter shank placed across the animal's haunches, gently force him up the ramp. The two most gentle animals should be the ones loaded first and last. To unload, the animals are turned, if possible, and led head first down the ramp. In some cases where animals are loaded in the direction of travel it is impossible to turn the first one to be unloaded so that he can be led. In this case he must be carefully backed down the ramp.

e. Care of animals during motor transport.—(1) Most animals soon become accustomed to loading, motor movements, and unloading and they can be handled without great difficulty. However, during extended trips they will suffer nervous fatigue and should be unloaded daily for feeding and watering. If the bodies of the trucks permit the attachment of improvised hayracks so animals can eat hay while en route, this will be found to be helpful.

(2) An unshod horse can retain his footing in a motor vehicle better than one with shoes; smooth-shod horses, however, may be transported provided floors are properly sanded. Motor transportation of animals shod with long or sharp calks is inadvisable unless individual stalls are provided in the vehicles used.

(3) Overhead cover to protect animals from sun and rain is advisable and can often be improvised by stretching paulins over tops of the vehicles not equipped with permanent covers.

CHAPTER 4

WAGON TRANSPORTATION

SECTION I

HARNESS AND HARNESSING

168. To put on and adjust halter.—The wagoner folds or coils the halter shank or rope, takes the halter and shank in the left hand, and approaches the animal just in front of the left shoulder. He places the halter under the animal's neck, grasps the crown piece with the right hand over the neck, slides it forward with the right hand, and extends the noseband forward with the left hand until it is in position; the halter is then raised by the use of both hands until the noseband is in position around the animal's nose. The crown piece of the halter is then buckled sufficiently tight to prevent its slipping off, but not so tight as to interfere with breathing, and its end is passed through the keeper.

169. To put on, adjust, and take off bridle.—*a*. To put on the bridle the wagoner takes the reins in the right hand and the crown piece in the left hand, approaches the animal on the left side, slips the reins over its head, and lets them rest on the neck. He then takes the crown piece in the right hand and places it in front of and just a little below its proper position. With the bit lying in the palm of the left hand, which is held under the animal's jaw, the thumb is inserted into the side of the mouth above the tusk. The wagoner presses open the lower jaw and inserts the bit by raising the crown piece. With the left hand the ears are gently drawn under the crown piece, beginning with the left ear, after which the throatlatch is secured.

b. The throatlatch should be loose enough to permit easy breathing and tight enough to hold the bridle securely on the animal's head. The cheekstraps should be adjusted so that the bit is well up in the corners of the mouth but does not wrinkle it.

c. To unbridle, the wagoner passes the reins over the animal's head, places them in the bends of the left arm, unbuckles the throatlatch, grasps the crown piece with the right hand, and gently disengages the ears. He removes the bit by lowering the crown piece.

170. To put on harness.—*a. Ambulance or escort wagon wheel.*—The wagoner first takes the breeching over his left arm, top down. He slips his right arm through the breeching and backband, grasps the right hame with his right hand at about the center and the left hame with his left hand, walks to the left side of the animal, and gently places the harness on its back. He now puts on the collar and adjusts the hames upon it. He pulls the breeching into place, the belly band through the loop in the end of the chokestrap, buckles it, snaps the side straps into the chokestrap ring and puts on the bridle.

b. Ambulance or escort wagon lead.—The wagoner proceeds as with the wheel harness, except that he places, adjusts, and secures the crupper in lieu of the breeching.

c. Buckboard.—The wagoner takes the bridle from the peg and passes it over his left arm, takes down the collar and places it on the same arm, takes down the harness with breeching and backband over his right arm, grasps the right hame about the center with his right hand and the left hame with his left hand, approaches the animal on the left side, gently places the harness on the animal's back, puts on the collar, and adjusts and fastens the hames upon it. He then draws the breeching into place and buckles the crupper around the tail, slips the end of the chokestrap over the belly band, buckles the band, and puts on the bridle.

d. Cart.—The wagoner takes down the harness and places the breeching over the right arm. Grasping the pad with the right hand on its rear end and the left hand on its front end, he approaches the animal on the left side and gently places the pad on its back. The breeching is then adjusted and the

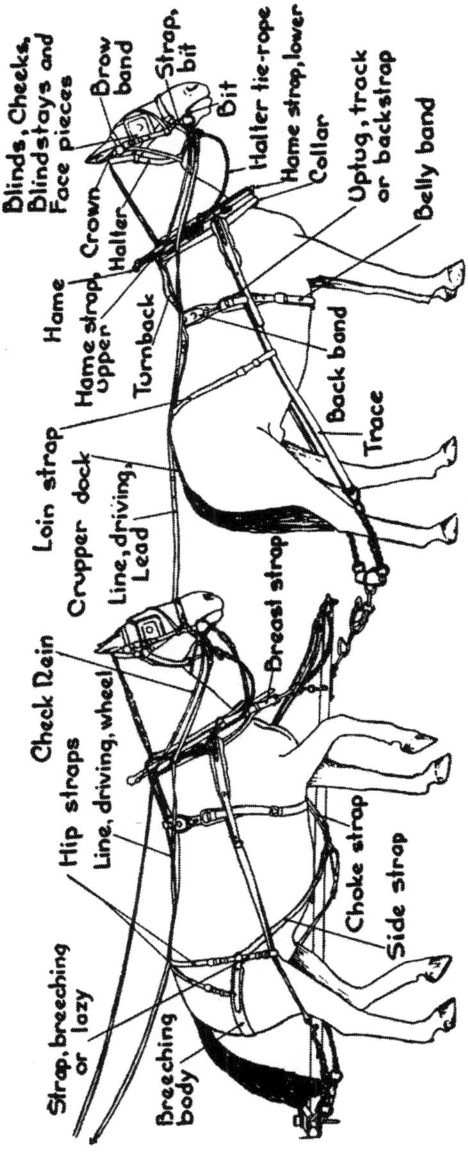

FIGURE 18.—Nomenclature of harness, ambulance or escort wagon.

belly band fastened. The wagoner then takes down the collar, places it on his left arm, takes down the hames, the right one in the right hand, the left one in the left hand, and places them on the animal's back in front of the pad and then puts on the collar. He then fastens the hames on the collar and puts on the bridle.

171. To adjust harness.—In adjusting various parts of harness, care should be taken that straps are not twisted, causing the buckles to be turned in toward the skin of the animal thus producing irritation and abrasions.

a. Collar.—A collar should be selected which when buckled will fit so that there is just room to pass the hand between the collar and throat at the bottom. It should not be so narrow that it will pinch the animal's neck or so wide that the bearing surface will come on the outer part of the shoulder. The top of the collar should be so fitted as to follow the contour of the crest and conform to the animal's neck. If the collar is too short it will choke the animal; if too long or too wide it will cause abrasions on the point of the shoulders. The collar should be moist when first used so that it will shape itself to the animal's shoulders and neck by the pressure of draft. This moisture is most effectively given by wrapping the collar in wet sacks and leaving them on for one night. Collar pads are useful with large collars when proper sizes cannot be obtained.

b. Hames.—The hames should be adjusted in length by shifting the upper straps to the proper holes. The length should be such that the point of the attachment of the trace will be opposite the middle of the surface of the shoulder over which the collar rests. This distributes the pressure evenly over the shoulder. In adjusting hames, the upper hame strap should be just long enough to be taut when the lower strap is buckled snugly.

c. Trace.—The traces should be adjusted to the larger animal of the pair. All four traces of the pair should be of equal length. The traces are hooked long enough so that the singletrees will not interfere with the movement of the animals; the wheel traces so adjusted as to allow at least 14 inches between the hindquarters and the singletree when in draft.

d. Breast strap.—The breast strap should be so adjusted that it will not be tight or leave too much slack to guide the pole properly. This strap is not intended to hold back the load.

e. Breeching.—The turnback strap should be adjusted so that the ring in the end will be even with the point of the hips, and the hip straps should be adjusted so that the breeching will hang squarely around the buttocks. A breeching hung too low has a tendency to pull the animal's feet from under him when backing. The side straps should be adjusted so that there is no strain upon the breeching other than when backing the vehicle and so that they do not have an unnecessary amount of slack at other times.

f. Lines.—The spread lines are passed through the rings on the spreader straps and adjusted so that the animals will be squarely in front of their own singletrees. If the spread lines are too short they will draw the animals' heads together; if too long they will allow their heads to spread apart.

g. Backband.—The backband is adjusted so that there will be no slack above the traces and no downward strain upon it from the traces when the animal is in draft.

h. Belly band.—The belly band should be buckled sufficiently tight to hold the harness in position but loose enough to permit the traces straightening without bringing a strain upon it.

i. Crupper.—The turnback should be adjusted so that the crupper will not pull on the animal's tail or be allowed to hang down from its proper position. The hip straps should be adjusted to such length that there will be no strain upon them when the animal is in draft and sufficiently short to have no unnecessary or unsightly slack.

172. To prevent and care for bunches.—*a.* Careful fitting, adjusting, and cleaning of harness and equipment and care in use of the lines and brake, particularly at the time of starting heavy loads, will prevent bunches. Prompt attention will prevent bunches from becoming large. It is much easier to prevent them than to eradicate those already formed.

b. Padding should never be placed over a bunch, as it only makes the bunch worse. If felt collar pads are available, the padding should be removed from them so as to relieve pressure from the bunch. If felt collar pads are not obtainable, pieces of sheepskin with thongs attached may be tied on the collar on each side of the affected part, and covering the entire half of the collar except over the sore spot. Small pieces would only cause undue pressure on the spots covered by them.

173. To unharness.—*a. Ambulance or escort wagon.*—The wagoner unbridles and hangs the bridle on the left hame by one of the bit rings or by the crown piece. He then unfastens the side straps from the chokestrap ring, unbuckles the belly band and slips it out of the chokestrap loop, unbuckles the lower hame strap, raises the hames, and places them in rear of the collar, unbuckles the collar and places it over the right arm. He then slides the harness over the animal's croup, passes the right arm through the breeching and backband, and grasps the right hame with the right hand about the center and the left hame with the left hand. He then hangs the hames on the harness peg by the upper hame strap, allows the collar to fall from the arm down over the turnback strap, and hangs the breeching on the harness peg. Lead harness is taken off in a similar manner, except that the crupper is hung on the harness peg in lieu of the breeching.

b. Buckboard.—The wagoner unbridles, slides the left arm through the bridle so that the reins and crown piece hang upon it, unfastens the lower hame strap, raises the hames and places them in rear of the collar, unbuckles the collar, takes it on the left arm with the bridle, and unfastens the belly band. He slides the harness off over the animal's croup, allowing the breeching and backband to pass over the right arm, and grasping the hames, one in each hand near the center, hangs hames, backband, breeching, collar, and bridle on the harness peg.

c. Cart.—The wagoner unbridles and unbuckles the lower hame strap and the collar. Taking the collar on his left arm and the hames in both hands, he hangs the hames and collar on the harness peg. He then unbuckles the belly band, slides the saddle off over the animal's croup, and places it on the harness peg, hanging the breeching over it.

174. To hitch.—*a. Team consisting of two pairs.*—(1) (*a*) *First method.*—The animals being harnessed and in their stalls, the wagoner backs out the wheel team and snaps the breast strap of the left wheeler into the bit ring of the right wheeler. He then backs the lead team out of its stall into position in front of the wheelers, snaps the spread line from the left leader into the bit ring of the right leader, hitches the left trace of the left leader into the bit ring of the left wheeler, and takes the team out to position in front of the wagon by leading the left leader. Upon arriving in front of the vehicle tongue, he fastens the other lead spread line and takes down the left leader's line which he holds in one hand, releases the trace from the left wheeler's bit ring, backs the wheelers into place, passes the left lead line through the ring on left wheeler's rein strap and places it over the left wheeler's back, fastens the spread lines on the wheelers, snapping them in the bit ring of the opposite animal. He then passes the breast straps through the rings of the pole chains and snaps them into opposite hame rings and takes down the left wheeler's line. He steps behind the left wheeler and hooks the left trace on the right wheeler, right trace on left wheeler, left trace on left wheeler, passes the lines over to the right side of wheelers, hooks left trace on left leader, right trace on left leader, left trace on right leader, right trace on right leader, takes down right leader's line and passes it through the ring on right wheeler's rein strap, takes down the line on right wheeler and hooks right trace on right wheeler, all in order named, after which he gathers up the lines and mounts to the seat.

(*b*) *Second method.*—The animals being harnessed and in their stalls, the wagoner enters the stall on left side of left wheeler, unties and passes in front of it, unties right animal, and standing at their heads, between and facing them, grasps both bridle reins near the bit, right hand holding left animal and left hand holding right animal, and backs them out of their stall. He then releases the outside rein and, leading them, walks between them, causing them to turn so that the left animal will be on the right side. He then brings out the lead team in like manner, places each in rear of its wheeler, and ties it by its halter shank to its wheeler's backstrap. He then takes the team to the vehicle by leading the wheelers, approaches the vehicle from the front, stopping at the front end of the pole and causes the wheelers to turn in alongside of it. He then snaps the inside line of left wheeler to right wheeler and the inside line of right wheeler to left wheeler, takes down the breast straps and fastens them through the pole-chain rings and attaches them to opposite hame rings, steps to right side and takes down right wheeler's line, hooks the right trace, passes around head of the team, takes down left wheeler's line, steps behind left animal, hooks inside trace of right animal and inside and outside trace of left animal, unties left leader and then right leader and leads them to position in front of the wheelers, attaches spread lines, takes down left leader's line, passes it through the ring on left wheeler's rein strap and places it on the left wheeler's back, hooks inside trace of right leader, inside and outside trace of left leader, takes up lines of the two wheelers and left leader and passes them over to the right side of the wheelers, takes down right leader's line, passes it through the ring on right wheeler's rein strap and places it with the other lines. He then hooks the right leader's right trace, gathers up the lines, and mounts to the seat.

(2) Either of the above methods may be used, the first being the most expeditious with animals which are docile and will stand quietly. The second method is preferable for animals which will not stand quietly. Either method when used with untrained or nervous animals will require the assistance of one or more men to hold each pair while the wagoner hitches them.

b. Team consisting of a single pair.—The wagoner backs the pair from their stall and leading them, passes between them so as to have the right animal on the left side, takes them to the vehicle by leading one animal with each hand, approaches it from the front, and leads them to a point opposite the middle of the tongue, one being on each side of it, turns them about and into position, fastens the spread lines and breast straps, takes down line on left animal and tosses it over to the left side of right animal, hooks left trace on left animal, passes around in front of the pair, takes down line on right animal, hooks right trace on left animal, left trace on right animal, right trace on right animal, gathers up the lines, and mounts to seat. One man can ordinarily hitch a single pair without assistance, but with untrained or nervous animals it may be necessary for an assistant to stand in front of the pair until the hitching is completed.

c. Single animal.—(1) *To cart with cart harness.*—The wagoner leads out the animal and backs it in between the shafts, takes down the lines, raises the shafts into position and passes the chain over the saddle, fastening it to the opposite shaft. He then hooks the trace chains into the hooks on the hames, holdback chains into the hooks on the shafts and fastens the bellyband, after which he takes the lines and mounts to the seat.

(2) *To other vehicles with light express or single buckboard harness.*—The animal is backed out of the stall and let to a position in front of and facing from the vehicle. The lines are taken down and fastened into the bit rings. If the vehicle is a light one, it is then run forward into position with one of the shafts on each side of the animal. If it is a heavy vehicle the animal is backed into position between the shafts which are run through the shaft tugs. The wagoner then takes down the line on the right side and passes it through the ring on the hame and through the terret, hooks the right trace and wraps the backing strap once or twice around the shaft from the bottom up, passes the end through the loop on the under side of the shaft and buckles it. He then passes it to the left side, takes down the terret, hooks the left trace, fastens the backing strap as explained for other side, buckles the belly band, hooks the check rein, takes the lines, and mounts to the seat.

175. To unhitch.—*a. Team consisting of two pairs.*—(1) (*a*) *First method.*—The wagoner unhooks the right trace on right wheeler, left trace on right wheeler, right trace on left wheeler, does up line on right wheeler, unhooks right trace on right leader, left trace on right leader, right trace on left leader, does up line on right leader, unsnaps spread line from bit ring on left leader, does up line on left leader, unhooks left trace on left leader, unhooks breast straps on wheelers, snapping breast strap on left wheeler into bit ring of right wheeler, unsnaps spread lines on wheelers, does up line on left wheeler, unhooks left trace on left wheeler, and hooks trace from left leader into bit ring of left wheeler. He then takes the team to the stable by leading the left leader. He unhooks the trace from the bit ring of left leader, unsnaps spread line from the bit ring of the right leader, and leads the lead pair into its stall, after which he unsnaps the breast straps from the bit ring of the right wheeler and leads the wheel pair into its stall.

(*b*) *Second method.*—This differs from the first method in that all traces and the breast straps are unhooked before any of the lines are done up. The leaders are tied to the wheelers on their side by means of attaching the halter shank to the backstrap, and the team is taken to the stable by leading the wheelers.

(2) Either of the above methods may be used, the first being the most expeditious with animals which are docile and will stand quietly. The second method is preferable for animals which will not stand quietly. Either method when used with untrained or nervous animals will require the assistance of one or more men to hold each pair while the wagoner unhitches them.

b. Team consisting of a single pair.—The wagoner unhooks right trace on right animal, left trace on right animal, right trace on left animal, and does up line on right animal. He then passes in front of the pair, unhooks left trace on left animal, does up line on left animal, unfastens pole straps, and unsnaps the spread line. He then unsnaps the outside lines from the bit rings and leads the pair to its stall.

c. Single animal.—(1) *From cart with cart harness.*—The wagoner unhooks hold-back chain on right side, unhooks trace chain on right side, does up lines on right side by folding them in a convenient number of folds and passing the bight through hame ring over hame and pulling them tight, unhooks hold-back chain on left side, unhooks trace on left side, unhooks chain holding up shafts, and lowers shafts to the ground. He then unsnaps the left line from the bit ring and leads the animal to its stall.

(2) *From other vehicles with light express or single buckboard harness.*—The wagoner unhitches and does up right trace, unfastens and does up backing strap on right side, does up the right line, unhooks and does up the left trace, unfastens and does up the backing strap on left side, unbuckles the belly band, does up the left line, slips the shaft tug off each shaft, and leads the animal to its stall.

d. To harness and hitch up by command.—For the purpose of instruction and to promote uniformity the various operations of harnessing and hitching up may be performed by number or command.

SECTION II

OPERATION

176. To drive.—*a.* The wagoner should sit comfortably in his seat, body erect, without stiffness, elbows close to the sides. He should not lean forward, half stand, or slouch back in his seat, because in these positions a falling or shying animal may pull him off his seat. The driving gloves should be large and comfortable. The hands should be held close together in front of the center of the body, knuckles to the front, forearms nearly horizontal.

b. In driving with the left hand, the right hand is used only for shortening the reins, assisting the left hand in stopping the team, and for using the whip. The left wrist should be slightly bent to the rear. This gives a more flexible contact with the animals' mouths than if the wrist is straight and rigid.

c. To hold the lines, the wagoner brings the line on the left wheeler under the little finger of his left hand up through the palm of his hand and over his thumb. He passes the line on the left leader between the little finger and the one next to it up through the palm of his hand and over his thumb. He brings the line from the right wheeler between the second and third fingers of the right hand and down inside the third and little fingers. He brings the line from the right leader between the first and second fingers of the right hand and down inside with the other line.

d. To hold all lines in one hand, the wagoner passes the lead line from the right hand over the top of the first finger of left hand and the wheel line from the right hand over the top of the second finger of left hand and turns both lines down inside the left hand.

e. In starting the team with a heavy load or in any situation where it is necessary to get the united power of the animals, the lines should be held fairly tight so that the animals may be made to feel the aid and guidance of the driver. In starting, all traces should be stretched before the team moves. A common fault is to start one pair before the others are in draft, the tendency of which is to make the animals balky, fatigue them by jerks, gall their shoulders, and break the harness.

f. While en route, the lines should be sufficiently tight to prevent their becoming entangled or an animal passing his tail over one of them, to give each animal the feeling of being on the hand, and to maintain pairs abreast of each other. In reducing the gait, halting, and backing, the pull on the lines should be gradual and used in conjunction with the voice and brake.

g. In turning to the right, the right lead and wheel reins should be grasped in front of the left hand by placing the middle and little fingers of the right hand over them. These reins should then be shortened by pulling on them, or the left reins lengthened by allowing the left hand to move slightly to the front. In turning to the left, the left lead and wheel reins should be handled in the same manner. In stopping, all four reins are shortened by pulling them through from behind, or the right hand may be placed over the reins in front of the left and the weight used to assist.

h. In crossing ruts or turning sharp corners the leaders should be out of draft; otherwise the pole may be snapped off or the wheelers pulled down.

i. Before arriving at the crest of a hill the team should be steadied for the descent, and with just enough traction from the leaders to prevent their singletrees hitting them. If a wheeler slips in descending the hill, especially if near the bottom, he should not be pulled up, but the team should be allowed to go a little faster.

j. In no case should the lines be used to accelerate the pace or gait, and the wagoner should never be permitted to "milk" the lines.

k. Jerking the lines should never be tolerated other than to stop a team which is running away.

177. Whip.—*a*. To use the whip, the wagoner passes the lines to his left hand, holding the whip by the butt end of the stock and the end of the lash between his forefinger and thumb. With a backward swing of the arm he releases the end of the lash, allowing it to swing back in a circular motion, and with a forward stroke the lash is brought down upon the animal desired.

b. The whip is used only for slow or lazy animals and in a hard pull, in which case a slight cracking of it is preferable and usually sufficient.

c. Use of the whip should be limited to necessity, and unnecessary cracking of whips should be prohibited.

178. Brake.—To put on the brake, the lever is pushed forward either with the hand or foot, and is held by the ratchet. To take off the brake, the catch is released from the ratchet and the lever is pulled back. The brake is used in going downhill, in crossing ditches, to stop the wagon, at a halt, and to hold it while the team is at a standstill. If a team starts to run away, the brake should be used to help stop it.

179. To release stalled wagons.—*a*. In pulling heavy loads or on heavy ground, if there is any liklihood of the team stopping, the wagoner should stop the animals before they get stalled. Many animals will not pull again on a load in the same place where they have met with resistance which has impressed them as insurmountable. If the team should get stuck, it may be induced to try again by turning the tongue slightly to the right or left and then getting all off together, straightening the pole as the pull is commenced.

b. If the wheels are sunk, the earth in front of them should be removed by using the spade prior to attempting to start, and if the wagon is in a very difficult position time will usually be gained by attaching additional animals, making use of the snatch block and fall, or unloading the wagon.

c. Should a wagon upset, the cover, bows, and lash rope are removed, after which it is unloaded, placed upright on good ground, and then reloaded.

180. To load wagons.—*a*. Animal-drawn vehicles should not be overloaded, as doing so will result in injury to them and to the harness and animals. The standard load of 3,000 pounds for the escort wagon includes 235 pounds for weight of the driver, his equipment, and the grain carried for the team. The standard load for the spring wagon is 1,000 pounds.

b. In loading, heavy articles should be placed in the bottom of wagon bed and in front, with lighter articles in rear and on top. Articles should be placed so that there is no vacant space between them. Articles which may be required en route should be so placed in the load as to be always accessible. The tail gate should be closed to prevent shifting of the load and to brace the sides of the bed. The bows should be pushed well down into their staples. The cover is then placed on the

134205°—39——12

bows and tied down to the rings on the side of the bed. Its ends are drawn together by a rope, the ends of which are then crossed and made fast to rings on the opposite sides of the wagon bed.

c. When wagons are loaded, the loads should be securely lashed with the escort wagon lash rope before the cover is put on. Ordinary loads are lashed by securing an end of the rope around one end of the front cross bar and lacing the rope from side to side over the load, using the lash cleats provided on the sides of the wagon bed for engaging the rope. The rope should be drawn tight and passed around one end of the rear cross bar, after which it is carried forward along the sides of the load by passing it around the first (rear) loop that goes over the load and taking all slack out by pulling it to the rear, after which it is carried to the next loop and so on until the forward loop is reached, to which it is attached by two or more half hitches after all slack has been taken out.

d. When possible, cargo vehicles should be loaded in the evening, after which they are placed under guard.

181. Record of cargo.—*a*. Cargo is checked as it is loaded on the wagons assigned to each section by its assistant wagon master and checked off at the destination by verification with the check slip used in loading.

b. A waybill should be prepared in triplicate by consolidating the loading slips for each wagon which should show the date, place of receipt, and place of delivery. The officer in charge receipts for the cargo by furnishing one copy of this waybill to the person from whom it is received, and upon delivery of the cargo gives one copy of the waybill to the person to whom it is delivered, taking receipt for it upon the other copy, which he retains.

182. Care of cargo and adjustment of loads on the march.—*a*. Special effort should be made by all personnel to care for and protect all cargo properly. Each wagoner will be responsible for his load while on the march and the guard responsible for all cargo while the company is parked.

b. Should the load of a wagon require attention on the march, an assistant wagon master of the section to which the vehicle pertains at once takes charge of it, causes it to be moved clear of the marching column and the load to be unlashed, replaced in position, and relashed, after which the vehicle resumes its march, taking position in any portion of the column which may be passing at that time. The assistant wagon master and the mounted personnel who have assisted in adjusting the load take up an increased gait and rejoin their section. The dismounted men who have assisted in adjusting the load are assigned to the nearest wagons, upon which they ride for the remainder of the day's march. In case of a breakdown, should it be possible to repair the broken wagon, the load is removed, if necessary, the repairs made, the wagon reloaded, and again placed in column. In case the wagon cannot be repaired, necessitating abandonment, it is unloaded, removed from the road, the load transferred to a spare wagon or distributed throughout the train, and the wagon destroyed if there is danger of its falling into the hands of the enemy.

c. In the case of draft animals which are sick, weak, or temporarily disabled, the team will halt clear of the column and the animal or animals will be replaced by extra ones. The team will then take position in the nearest section and rejoin its proper section at the end of the day's march.

183. Rates of travel.—The rates of travel of animal-drawn transportation vary with the nature of the vehicle, the size of the load, the class and condition of the animals employed, condition of the roads and weather, and length of the column. Passenger-carrying vehicles can be driven at the rate of 7 miles per hour for 3 or 4 hours under favorable conditions. Should the distance be great, arrangements should be made to change the animals every 15 miles by placing them in relays. Heavily loaded vehicles can maintain the rate of march prescribed for infantry under favorable conditions.

184. Road rules.—The rules in common use by drivers of private vehicles, no matter whether in the United States or foreign countries, will be strictly adhered to by drivers of all Government vehicles. Halts will always be made at one side of the road, leaving a clear passage. In the United States, Hawaii, Panama, and Puerto Rico the rule will be followed of passing on the right and overtaking on the left. On a hill, all teams coming up have the right-of-way over those going down. In case of a break-down the vehicle will be immediately removed so as not to obstruct traffic.

Section III

MAINTENANCE

185. Vehicles.—*a. To assemble.*—Animal-drawn vehicles are usually shipped knocked down, and their component parts should be carefully checked as soon as uncrated in order to be certain that all are present. A list of component parts with instructions for assembling will ordinarily accompany the shipment. A study of the details for assembling the particular vehicle should be made before attempting to assemble it. All nuts and bolts should be tightened at time of assembling and frequently thereafter, as the life of animal-drawn vehicles depends to a great extent upon keeping the nuts and bolts tight.

b. Care.—Vehicles require daily examination and repair.

(1) Upon return of troops from field duty the transportation should be immediately overhauled, repaired, and necessary accessories replaced.

(2) At each inspection of the transportation, required repairs should be noted and assigned to individuals whose duty it is to make them.

(3) When on field duty the transportation must be carefully examined at the end of each day's march to see that the spokes of wheels and nuts on axles and bolts are tight and to arrange for proper repairs of vehicles and harness, including the replacement of broken or lost spare parts.

(4) Tail gates and chains will not be allowed to hang down or drag.

(5) In garrison, or in camp where there is a probability of movement by rail, blocks for securing vehicles on freight cars and a ramp for unloading vehicles and animals should be prepared and kept on hand.

(6) Specifications and blue prints of animal-drawn vehicles will be furnished to persons concerned upon application to the commanding officer, Jeffersonville Quartermaster Depot, Jeffersonville, Indiana.

(7) Vehicles will be greased twice a week regularly and as often in addition as may be found necessary. If in constant use, wagons will require daily greasing. In greasing, the old grease, dirt, and sand should be removed. The axle is usually raised for greasing by use of a jack. Should it be necessary

to grease a heavy vehicle without use of a jack, the nut is first removed. A board or stick, slightly longer than the distance from the axle to the ground, is secured and placed with one end on the ground and inside the wheel, the other end leaning toward the wheel; the upper end is placed under the spoke which is about horizontal. The wheel is then raised slightly by lifting carefully on the spoke on the opposite side; when the spindle has been properly greased, the weight of the vehicle will cause the wheel to slide down the spindle. It should not be allowed to slide more than three-fourths of the way off, but should be taken off far enough to scrape off the old grease. This method cannot be used if the axles have not been properly greased. The wheel will not run on as easily as it did off, but with the stick inclined and a slight pressure on the wheel it will go on. The nearer the upper end of the stick is placed to the hub the less power will be required to lift the wheel off the ground.

(8) All vehicles in use should be painted olive-drab color at least once a year. The formula for mixing olive-drab paint is 6 pounds white lead ground in raw linseed oil, 1 pound raw umber, 1 pint turpentine, one-half pint japan dryer, and 1 quart raw linseed oil. Iron parts should be painted black by use of the following mixture: 1 pound drop black and 1½ pints coach varnish for each wagon. The canvas cover of stationary tops of light vehicles and ambulances should be waterproofed by use of the following mixture: 1 gallon boiled linseed oil, 1 pint dryer, and 1 pound dry ocher. Thoroughly mix the material and let it stand for 24 hours before use. Before applying this material the canvas surface should receive a sizing coat of water and soft soap as thin as paint. The paint is to be applied while the canvas is still wet and rubbed in brickly, using a 4-inch common paint brush.

186. Harness.—*a. Arrangement.*—(1) In garrison, harness should be conveniently arranged on pegs on the heel posts of stalls and, if practicable, should be sacked to keep off the dust; or it should be placed on proper supports in a harness room.

(2) In the field when no other shelter has been provided, harness should be kept in the wagon bed.

b. Care.—(1) When removed from the animals at night, harness should be thoroughly cleaned with saddle or harness soap and water. It should be oiled once a month, or more often if necessary. The harness should never be thrown on the ground where it will get muddy and dirty. Collars should be buckled when removed from the animals.

(2) To clean and oil black leather harness, take the harness completely apart, unbuckling straps, and removing buckles, loops, etc., wherever possible. The equipment to be used is a bucket of lukewarm water, sponge, harness soap, harness dressing, neat's-foot oil, and lampblack. Hot water must never be used or the leather allowed to soak in any water. A rack should be provided on which to hang the harness. When no better arrangement is available, one end of the wagon pole can be inserted between the spokes of one of the hind wheels, above the hub, and strapped to the axle. A set of harness can then be hung on the pole, the sponge moistened in clean water and passed over the harness until the dirt becomes soft The sponge should be rinsed as often as necessary and the dirty water replaced frequently with clean water. A good lather is then obtained by rubbing the sponge on the harness soap and the harness is given a good heavy coating of this lather, the rubbing being continued until all the dirt and lather are removed. In some instances it may be necessary to use a thin piece of wood to scrape off the dirt.

(3) When the harness has not been cleaned for some time and is hard it should be cleaned as described in (2) above, then oiled. For each single set of harness to be oiled, a pint of neat's-foot oil is poured into a pan, mixed with lampblack in the proportion of 1 teaspoonful to each pint of oil, and stirred until it has a glossy black appearance. In cold weather the oil should be heated until it is lukewarm, but never hot, before using on harness. The mixture is applied with a small sponge, rubbing it in well. About 48 hours should elapse before using harness again. In cold weather the harness after being thus oiled should be allowed to hang in a warm room for an hour. It must be dried gradually; never close to a hot fire or stove or in the hot sun.

(4) In cleaning and oiling fair leather harness the procedure is the same, omitting the use of lampblack and such soap as tends to blacken the leather.

(5) A buckle must not be worn in the same hole all the time, but should be shifted a hole or so occasionally, as this precaution will prevent the leather from rotting in that place.

(6) The wagon master should examine the harness at each inspection of transportation, and each driver should examine that of his own team every day upon completion of work. The necessary repairs should be made without delay.

(7) Cleaning and preserving materials for equipment are furnished by the Quartermaster Corps.

Section IV

MARCHES, DISCIPLINE, AND SIGNALS

187. Marches.—*a.* All animal-drawn transportation not otherwise employed should make practice marches of from 5 to 10 miles per day, Sundays and holidays excepted, in order to keep personnel, animals, and equipment in condition. During these marches the personnel are instructed in their duties on the march, in bivouac, and in camp, and the animals are grazed for short periods when practicable.

b. The normal formation for marching on ordinary roads is column of wagons, the officer in charge and wagon master at the head of the column. If the number of wagons in the train exceeds 20, organization into two or more sections should be made. Each section should be commanded and led by an experienced noncommissioned officer, mounted on a horse. In addition, two or more mounted men should be assigned to each section to assist in maintaining march discipline, to pass on signals, and to assist the train over difficult places. Section leaders ride at the head of their sections, one mounted man at the rear. Other personnel individually mounted are spaced throughout the column, each one observing the wagons forward of him and in rear of the man he follows. The normal distance between sections is 20 yards. The positions of sections in the column should be rotated daily. From time to time section leaders halt on the flank of the column and observe the wagons of their units as they pass, resuming their normal positions after observing the last vehicle. Like observations of the entire train are made from time to time during the march by the train commander and wagon master, who alternate such duty so that one of them is always at the head of the column during the march. These observations should determine the condition of animals, vehicles, and loads, whether any animals are in distress, any loads require adjustment, whether teams are being driven so that each animal is doing its share of the work, and if proper distances are being maintained. A mounted man should be stationed near each difficult place to warn each wagoner prior to his approaching it. This man resumes his former position after the last wagon has passed.

c. In forming column for a march, the teams should be placed according to their relative speed at the walk, the faster-walking teams in the lead.

d. In ascending and descending steep slopes, the column should be halted and wagons sent up at short intervals, which will give opportunity for each team to rest at the top of the hill. When halted a number of times during an ascent, each team is moved out in time for the next one to halt on the same ground. At such halts the brake should be put on firmly, and if insufficient to remove the strain from the animals the hind wheels should be blocked. In descending, if application of the brake is insufficient to hold the wagon off the team, the lock chain should be used on one or both hind wheels.

e. During forced marches the loads should be light and should not ordinarily exceed 2,000 pounds. Night marches are difficult for wagon columns, although animals usually travel well at night. If the night is dark, proper observation and supervision are nearly impossible. Wagons must be kept well closed up, all personnel especially alert, special care taken to prevent the head of the column from taking the wrong road, the entire column halted if any loads require adjustment, and lights provided to aid wagoners while crossing difficult places.

f. Difficult crossings must be reconnoitered before being attempted. If the road is such that portions of it must be repaired, several men should be placed on wagons near the head of the column with picks, spades, and axes for the hasty repairing of such places. This may be done by cutting brushwood, placing it across the road, throwing earth upon it, and then tramping same down. Should a stream requiring fording be encountered, the column will halt and a mounted man sent across to ascertain the depth, velocity of the current, and condition of the bottom and approaches, after which the wagons will be sent across with distances of about 30 yards between them. If the stream is so deep as to be dangerous, mounted men will be stationed along the down-stream side. If deep enough to require swimming, the officer in charge should seek another crossing or arrange for having the vehicles and loads crossed in boats or on rafts, the harness being placed on the vehicles, and the animals crossed by swimming. Should the approaches be difficult, they should be prepared before crossing is attempted, a few of the leading wagons halted at the side of the road, and the teams used as "snatch" teams for doubling the number of animals on each vehicle during the time of its crossing. In crossing weak bridges, wagons should proceed at the walk with increased distances.

g. An escort wagon may be crossed by removing the wagon bed from the running gear, placing it in the canvas wagon cover or a large paulin, and drawing the cover or paulin taut, thus improvising a boat. The wagon bed is then lashed to the running gear and rolled into the stream. Some of the wagonload may be placed in the wagon bed, the amount depending on the wind and current of the stream.

188. Road discipline.—*a.* Upon completion of hitching and loading, wagoners take position on the seats of their vehicles in readiness to move forward when directed to do so. There should be no driving about by individual teams unless same is directed by proper authority.

b. Each wagon will move forward in time so as to take its proper position in normal march order of single column as it leaves the camp.

c. The normal distance from the head of one team to the tail gate of the preceding wagon is 5 yards, which distance will be maintained unless otherwise directed.

d. At halts, wagoners will immediately make an examination of their team, harness, vehicle, and load, making any adjustments required, and then resume their position on the seat. Men individually mounted will dismount and examine their animals and equipment, retaining their position in the column and mounting as it moves forward. After examination of the animals and adjustment of equipment, all personnel direct their attention to the head of the column in order that they may observe the signal for resuming the march.

189. Transportation signals.—*a.* To avoid noise and confusion, signals with the arm or hand or headdress may be used when they can be clearly seen. Every signal should be made with vigor and precision, and every position taken by the hand and arm during the signal should be distinct and clearcut. The whistle blast may be used to attract attention preparatory to giving a signal. Section leaders repeat the signals of the column commander or make appropriate signals of their own.

b. The signals prescribed below are preparatory signals unless otherwise indicated; for the signal of execution the arm is extended vertically at the conclusion of the preparatory signal, and then lowered quickly to the side.

(1) *Attention.*—Extend the hand upward vertically to the full extent of the arm, palm to the front, and move the arm and hand slowly back and forth several times from right to left, left to right, in a vertical plane. This signal may precede any command; not a preparatory signal.

(2) *Forward.*—Extend the hand vertically to the full extent of the arm, palm to the front, and lower the arm and hand in the direction of movement until horizontal.

(3) *Halt.*—Extend the hand vertically to the full extent of the arm.

(4) *Column.*—Raise the arm vertically, then drop the arm to the rear and describe circles on the right in a vertical plane with the arm fully extended.

(5) *Line.*—Raise the hand vertically to the full extent of the arm and wave the arm well down alternately to the right and left several times.

(6) *Change of direction.*—Extend the arm horizontally, with the palm down, in the direction of the marching flank, and then make a wide sweeping movement with the arm in a horizontal plane toward the new direction.

(7) *Left (right) front into line.*—Signal line, and point in the direction in which the line is to extend.

(8) *Disregard.*—Place the hand against the back at the height of the waist, back of the hand toward the body; not a preparatory signal.

(9) *Increase the gait.*—Carry the hand to the shoulder, forearm vertical; extend the arm vertically from this position and repeat several times.

(10) *Decrease the gait.*—Hold the forearm horizontally above and in front of the forehead.

SECTION V

BY RAIL AND WATER

190. Rail.—*a. To entrain.*—(1) *General rules.*—All quartermasters should familiarize themselves with instructions covering troop movements contained in AR 30–930, 30–935, 30–940, 30–945, and 30–965.

(*a*) Escort wagons, animal-drawn ambulances, and other similar nonmotorized vehicles should always be shipped "knocked down," unless it is anticipated that conditions will not permit the delay necessary to reassemble such vehicles upon arrival at destination.

(*b*) Care should be exercised to discover in advance of the shipment whether or not proper facilities for unloading exist at the detraining point.

(*c*) Where conditions do not permit shipping of vehicles "knocked down" they are shipped on their own wheels. The drop-end gondola is the most suitable car for all types of animal-drawn vehicles which are shipped on their own wheels, including nonmotorized artillery, engineer, and signal types.

(*d*) A permanent loading detail consisting of experienced officers and men, the number depending upon the size of the movement, should be selected at the entraining point to load all mobile equipment. This detail should be subdivided into teams for dropping brakes and gondola ends, repairing car floors, driving in nails, hauling vehicles on to cars, blocking and lashing, and brake replacing.

(*e*) The above details are separate and distinct from those required for loading of company or troop property.

(*f*) Organizations should deliver vehicles at entraining points in proper condition for shipment. Tops, tail gates, etc., should be securely fastened to prevent wind whip or loss and damage.

(2) *To load from ramps.*—Animal-drawn vehicles can be easily and rapidly loaded from permanent or temporary end

ramps on flat cars or drop-end gondolas, six to ten cars in a string, either by hand or mechanical towing, if runways made of 2 inch by 12 inch by 8 foot planks, ends beveled, with 2 by 4 inch strips on the bottom to prevent slipping are placed between cars. The first vehicle loaded should be pulled through to its place on the last car and the others following. As vehicles are loaded the runways may be removed and spiked to car floor for use in unloading.

(3) *Construction of temporary ramp* (*end ramp*).—Temporary end ramps may be constructed in a number of ways, depending upon material available.

(*a*) A cribwork may be built up at end of cars to be loaded, using railroad ties or similar timber. On this crib nail four 2 inch by 12 inch by 12 foot planks. A ramp of this kind will hold any vehicle except the heavy guns.

(*b*) Two pairs of 8 inch by 12 inch by 16 foot timbers, with ends beveled, make a good temporary ramp when properly supported by cribwork.

(*c*) Two escort wagon wheels with axle and two planks may be used as a temporary end ramp.

(4) *To block.*—All vehicles should be properly blocked and secured before train is permitted to depart. For escort wagons, animal-drawn ambulances, spring wagons, carts, rolling kitchens, and vehicles of a similar nature, 2- by 4-inch lumber may be used as follows: Two pieces of 2- by 4-inch lumber about 3 feet long should be spiked to car floor, parallel with each wheel, one piece on the inside of wheel and the other on the outside. To these pieces a 2- by 4-inch piece about 8 feet long should be spiked in front and rear of each wheel. A short piece should be spiked across each set of 2- by 4-inch pieces to

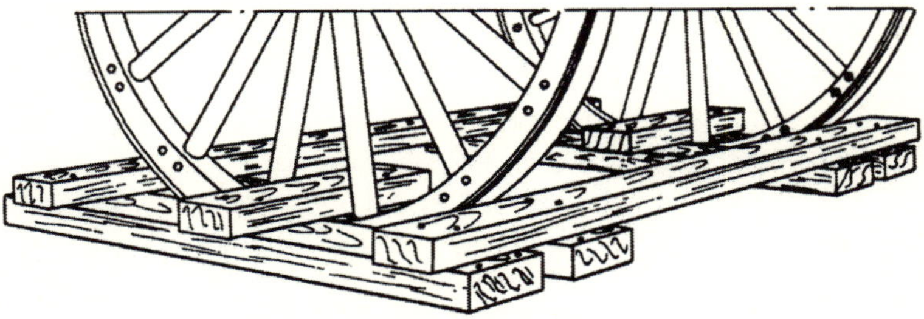

FIGURE 19.—Wagon block.

prevent wheel from slipping out. A further precaution may be taken by passing a piece of burlap or canvas around the felloe of the wheel and spiking it to the car floor.

b. To detrain.—Upon arrival of train at detraining point, if ramp has been constructed, have train moved so that the first car containing the first vehicles loaded is placed opposite the ramp, remove blocks from wheels of vehicles, place runways of 2-inch by 12-inch by 8-foot planks, ends beveled, between the cars and then proceed to pull the vehicles either by hand or mechanical means, if available, from the cars.

191. Water.—*a. To embark.*—(1) Animal-drawn vehicles for shipment by water may be grouped under two general classes; those vehicles which are shipped direct from a depot "knocked down" and consigned to a supply agency overseas, and those vehicles belonging to organizations shipped either with the organization or on a separate ship for delivery to the organization upon their arrival overseas.

(2) In all shipments timely notice is given the general superintendent, Army Transport Service, to prevent unnecessary accumulation of supplies awaiting shipment.

(3) Where shipments are to be made from a depot direct to a supply agency overseas the manner of handling the shipment is the same as for shipping ordinary heavy freight. Vehicles are shipped on a Government bill of lading consigned to the port from which shipment is to be made for transshipment to consignee. The bill of lading is accomplished at the port where vehicles are consigned and turned over to the last carrier. Cargo is loaded, discharged, and checked on and off the ships by checkers and stevedores under direction of the general superintendent, Army Transport Service, at ports where there are such officers; otherwise, under arrangements made by a transport quartermaster.

(4) In the case of animal-drawn vehicles belonging to organizations being transported on the same ship with their organization where vehicles are "knocked down", the manner of making the shipment is the same as outlined for shipment from depots. Where the necessity exists for immediate use upon debarking of certain vehicles, it will be necessary to ship these vehicles on their wheels. In such cases the number of vehicles to be shipped, together with their cubic dimensions, is

sent in advance to the general superintendent, Army Transport Service, at the port from which shipment is to be made in order that proper space on the ship may be assigned.

(5) In "unit shipments," such as referred to above, the commanding officer of the port will prepare the schedules of embarkation of troops, material, and supplies on transports after consultation with the commander of troops to be embarked. Following this, the commander of troops issues an embarkation order based upon this schedule, which prescribes the time at which vehicles will be placed on the docks ready for releasing to the transport service.

(6) Packing lists showing description, weight, and cubic dimensions of articles must be submitted by organization commander to the general superintendent, Army Transport Service. Organization commanders certify on this list that no unauthorized or inflammable materials are included in the organization property, except those prescribed by Army Regulations, which are separately boxed and plainly marked.

(7) Boxing, crating, etc., will be in accordance with standard War Department specifications for export shipments and will be marked in accordance with AR 850-20.

b. To debark.—(1) The discharge and checking out of all cargo is a function of the checkers and stevedores under the general superintendent, Army Transport Service. Responsibility of organization commanders begins when articles are unloaded at debarking points and checked to the organization commander.

(2) For further details see AR 30-930, 30-935, 30-940, 30-945, 30-1105, 30-1255, and 30-1285.

CHAPTER 5

PACK TRANSPORTATION

Section I

GENERAL

192. Classification.—*a.* Pack transportation is of three distinct types:

(1) Cargo pack trains, operated by the Quartermaster Corps and by pack artillery units. The loads, generally bulky and of considerable weight, are secured to the saddle by ropes. The gaits of the animals are the walk and amble.

(2) Artillery combat pack units with loads packed in steel hangers and carriers. The gaits of these animals are the walk and the amble.

(3) Cavalry pack transportation, with loads of reduced weight and bulk packed in hangers and carriers and so positioned as to enable the animals to maintain equilibrium at the walk, trot, and gallop. Special cargo loads are secured to the saddle with ropes. Infantry pack transport is placed in this classification.

b. The first two types use mules exclusively, while the third type may use either mules or horses. Each type requires skilled personnel to maintain pack transport in continuous operation.

193. Phillips pack saddles and equipment.—*a. Types of saddles.*—There are two types of the Phillips pack saddle. The smaller type is the cavalry pack saddle used for all cavalry and most infantry pack loads. The pads of this saddle are 22 by 19 inches and its weight is 43 pounds, exclusive of breeching, breast collar, and cinches. The larger type is the cargo pack saddle, used for all pack artillery loads, all pack train loads and the heavier infantry weapon loads. The pads of this saddle are 23 by 25 inches and its weight is 72 pounds, exclusive of breeching and cinches.

b. Equipment.—The saddle consists of a metal frame to which pads are attached. The saddle is equipped with specially designed breeching, breast collar, and cinches.

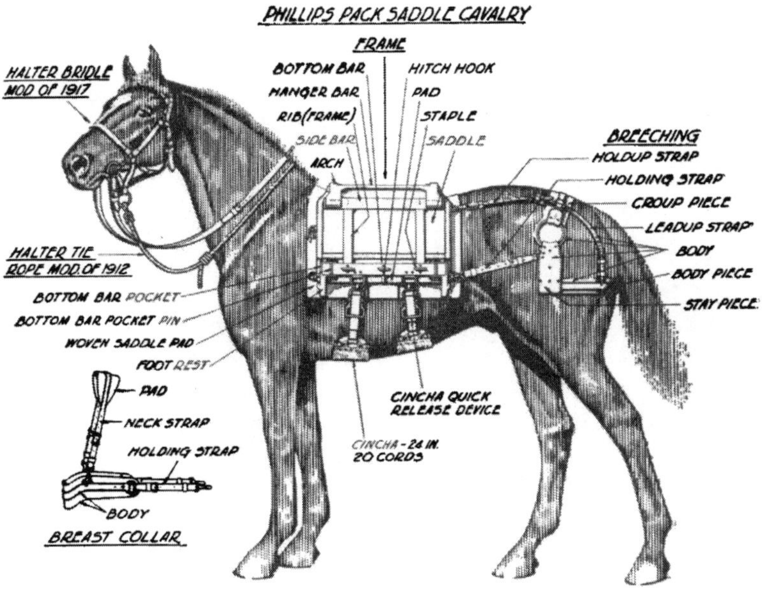

FIGURE 20.—Phillips pack saddle, cavalry.

(1) *Frame.*—The frame of the cavalry pack saddle is of aluminum alloy with the exception of the hooks, rings, and ribs. The frame of the cargo type pack saddle has aluminum alloy side and bottom bars with steel arches and hanger bars. The ribs of both frames are of spring steel, each shaped to a definite radius. Pads are easily and quickly attached to, or detached from, the frame without the use of tools.

(2) *Pads.*—The outside of the pads is of tan leather, reinforced internally with aluminum alloy. On the outside lower corners are metal pockets, with foot rests, for locking the bottom of the pad to the frame. The foot rests are used for holding the ropes of hitched loads and for keeping the pads off the

ground in the field. The contact side of the pads is of woven felt. The pads are stuffed with curled hair which retains its resiliency indefinitely.

(3) *Saddle cover.*—A canvas cover with leather borders covers the pads of the saddle to protect them from rain and wear.

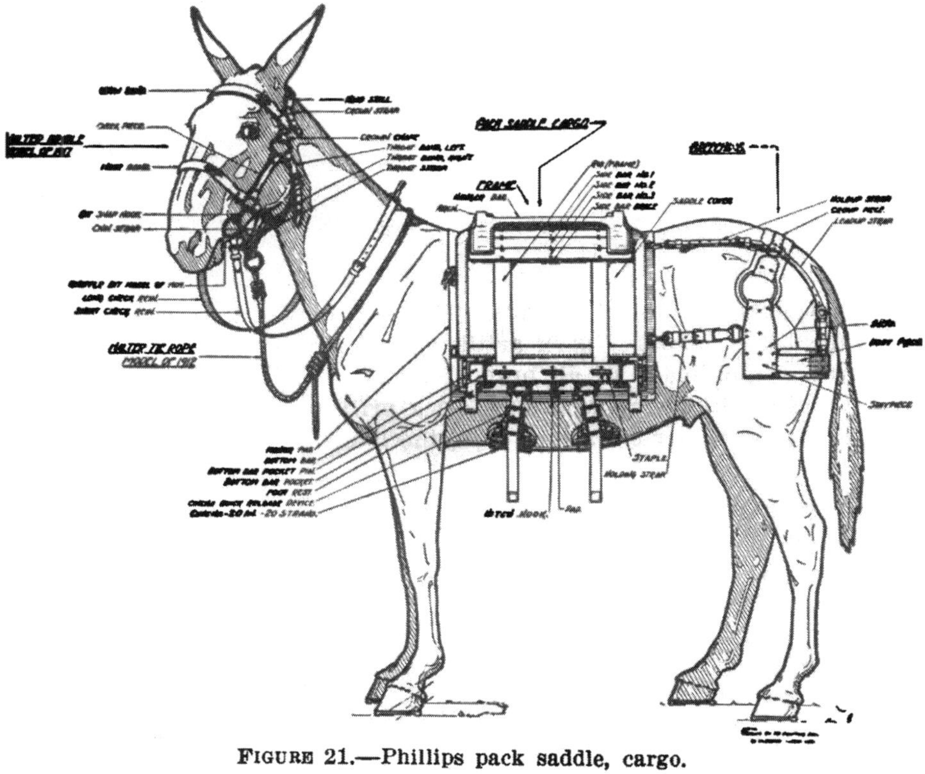

FIGURE 21.—Phillips pack saddle, cargo.

(4) *Cinches.*—The saddle is equipped with two adjustable mohair strand cinches. A cinching device is provided with each cinch. This device permits rapid cinching and release of cinches, careful adjustment of cinch pressure, and eliminates tying and untying knots.

(5) *Breeching.*—A leather breeching accompanies each saddle. This breeching functions principally through the two

lower holding straps that attach to the lower D-rings on the rear edges of the saddle pads.

(6) *Breast collar.*—The cavalry type saddle is normally equipped with a breast collar. The breast collar is useful for animals of poor conformation, at rapid gaits or on steep grades; it is not necessary for animals of normal conformation. The cargo pack saddle is not equipped with a breast collar.

(7) *Woven pad.*—This is a woven saddle pad that shapes itself instantly to the conformation of the animal; it is resilient and being porous it absorbs and evaporates sweat, thus keeping the animal's back cool. Leather thongs may be passed through the weave for attachment to the small rings on the front edges of pack saddle pads. Woven saddle pads for the cavalry type of pack saddle are 27 by 38 inches; for the cargo pack saddle, 27 by 44 inches.

(8) *Top load arches.*—Load arches of various types are provided for carrying top loads. These are of steel or bronze for the heavy loads of pack artillery and of aluminum alloy for the lighter loads of all branches. The Ordnance Department supplies the heavy arches for pack artillery and other combat loads; the Quartermaster Corps, the lighter top load arches for cargo loads.

c. Care and cleaning.—(1) *Saddle frame.*—No special precautions are required for the care of the saddle frame other than to keep all steel parts free of rust and all aluminum alloy parts free from corrosion by occasional painting of the frame. Neither strong alkalis nor acids should be used for removing paint from aluminum parts. However mild solutions of either may be used but all traces of same must be removed before applying paint. In case of a broken rib, a new spring steel rib from the spare parts kit should be substituted.

(2) *Saddle pads.*—All leather parts should be cleaned with saddle soap and sponge and occasionally receive a light coat of neat's-foot oil. The contact sides of the pads should be kept smooth and clean. All mud and the dried residue from sweat should be carefully removed by brushing.

(3) *Breeching and breast collar.*—All leather should be kept clean and free of grit and dirt. The contact parts of the breeching and breast collar must be free of any excess soap or oil which will collect dust and grit and thus rub the animal.

(4) *Woven saddle pads.*—Pads in storage should be protected from moths. In the field these pads should be kept smooth, soft, and free of grit. After unsaddling, the pads should be folded once with the wet side in and placed upon the pack saddle. Loose hair collected from the pack animal should be brushed out shortly after unsaddling and before it has had time to dry. The pads should receive a final brushing out several hours after unsaddling to free them of the salt from sweat. The pads should never be exposed to the sun to dry as this causes them to harden. They should be washed occasionally in cold water. They should always be inspected for cleanliness before saddling.

(5) *Girths.*—Mohair cord girths in storage should be protected from moths. In the field these girths should receive care similar to that applied to the woven saddle pad.

(6) *Cinching straps.*—Cinching straps should be kept clean and slightly oiled. When necessary to replace them, leather of the same thickness should be used and the new straps tested in the cinching device until they operate smoothly.

SECTION II

SELECTION AND TRAINING OF PACK ANIMALS

194. Selection.—*a.* Pack mules for artillery and quartermaster pack transportation are issued as such. Pack horses must be selected from animals issued as riding horses.

b. In general a pack horse should be from 15 to 15½ hands in height, and weigh from 1,000 to 1,150 pounds. He should be of compact, stocky build, with medium withers, straight, well-muscled back, deep girth, straight, large-boned legs, and good feet. Types of horses that cause trouble as pack horses are those with withers that are too thick, too flat, or too thin; back too short, too long, swayed, or roached; broad-ribbed, draft type of chest; and excessively large barrel.

c. In addition to desirable conformation, pack horses should be gentle and free of vicious habits, and should walk, trot, and gallop freely and boldly. There should be little motion to the back at the walk, and its action should be low at the trot. There must not be excessive transverse motion (side swaying) of the body in rear of the forequarters at the walk, nor excessive up and down movement of the saddle position at the trot.

d. Horses for use under pack are selected first from all the horses of the organization. When these have been selected and drivers assigned (if the pack horses are to be driven or led by troopers), the horses for the drivers are selected, an effort being made to pair pack and riding members of the team so as to get a smooth-working team.

195. Training.—*a.* The principles outlined in chapter 2 for the conditioning and training of remounts should be applied in the conditioning and training of pack animals. If practicable, all pack horses should be given basic remount training before their selection as pack animals.

b. When basic training has been completed, special training is given pack animals, including training to stand while being packed and unpacked. As conditioning and training progress, weight is gradually added and increasing distances are covered in practice marches, so that at the earliest practicable date pack animals are capable of covering average marches without undue fatigue or injury.

Section III

FITTING AND ADJUSTMENT OF EQUIPMENT

196. To fit the saddle.—*a.* The Phillips pack saddles fit a large percentage of animals without the necessity for special fitting or adjustment. For the other animals only minor adjustments will be required.

b. When new saddles are assigned to animals, the saddles should receive a breaking-in process before making any adjustments of the pads, except for those few animals whose conformation obviously requires special fitting. This breaking-in process removes the stiffness from the pads, and the saddles will then shape themselves to the conformation of the animals; places requiring adjustment will show up, and the adjustments may then be accurately made.

c. In breaking-in the saddle, it must be properly positioned on the animal's back and cinched. It is best broken-in by placing a balanced load on it and marching from 5 to 10 miles a day for a week or longer. A sack of oats placed across the saddle is satisfactory for this work, or a balanced side load may be used.

197. Saddling.—*a.* Too much stress cannot be laid upon the proper position of the saddle and upon correct cinch adjustment. The forward edge of the saddle should be sufficiently in rear of the shoulder blades for these members to operate. Two or three inches is the correct measurement. The saddle should not incline to a side, and the bottom bars should be horizontal or inclined slightly downward and forward, never downward to the rear.

b. Saddling should be practiced first without the breeching. The animals' back and sides should be groomed and the mohair pad clean and soft before saddling.

c. Standing on the left side of the animal, place the mohair pad upon the animal's back, short sides of the pad from front to rear, and smooth out all folds. The forward edge of the pad should be approximately 3 inches forward of the saddle when the saddle is in its proper position.

d. The saddle, with cinches attached to the right D-rings and resting on top of the saddle, should be about 1 yard in rear of the animal. The saddle is best handled by two men, one on each side taking hold of the foot rests and lifting it. The saddle is raised over the croup and placed on the animal's back in the correct position. It should be viewed from the rear to see that it is centered and from the sides to see that the bottom bars are horizontal or inclined slightly downward and forward.

e. With the saddle in position, the packer on the right side lowers the front cinch and passes it under the animal's belly to the packer on the left side, who grasps it and hooks the cinching device to the D of the saddle, takes up the slack of the running strap in the same manner as when cinching the McClellan saddle, and pulls the free end of the running strap until the metal loop is snug against the buckle of the device thus locking the strap. If necessary to lengthen the running strap in order to hook the cinching device to the cinch D of the saddle, push downward on the metal loop of the running strap while holding the device. When it is necessary to cinch very tightly, pull downward on the free end of the running strap until the required tension is obtained. This latter method, however, is seldom necessary and should not generally be used, as it has a tendency to pull the saddle off center toward the left side. For neatness after cinching, place the free end of the running strap through the metal loop and fold the strap back on itself. The rear cinch is then similarly lowered to place and cinched.

f. After cinching, observe the saddle from the rear to see that it is centered. If the saddle is inclined to one side place a forearm on the opposite bottom bar of the saddle and with the other hand pull the saddle to center.

g. The method of attachment and operation of the cinching devices described above is that used for the cavalry pack saddle. As the cargo pack saddle requires greater cinch pressure, especially on the front cinch, the cinching devices are attached to the cinch D's on the left side; the cinches on the right side. With the saddle properly positioned, the packer on the right side lowers the front cinch and passes it under the animal's belly to the packer on the left side, who grasps it with his left hand and with his right hand hooks the cinching device to the cinch D and takes up the slack of the device by pulling upward on its running strap. The metal loop of the device is drawn snug against the buckle to lock the straps. The rear cinch is then similarly lowered to place and cinched. Minimum pressure should be applied to the rear cinch.

h. To release the cinching device, pull outward on the metal loop to free the running strap, then grasp the buckle of the device at the top and pull outward and downward thus releasing the running strap, then detach the hook of the buckle from the cinch D.

198. Adjustment of cinches.—*a.* A pack saddle covers such a great amount of movable surface that proper cinching is vital to success. Excessive binding of the front cinch may injure the back or sides, interfere with breathing, or cause cinch galls. The rear cinch must not be as tight as the front cinch, for the rear of the saddle covers the area of greatest motion, the flexible, short-ribbed region and the tender region over the kidneys. The hindquarters oscillate from side to side in travel and have an up and down motion which must not be bound by cinch pressure. The rear cinch should be only sufficiently tight to hold the type of load carried and for the rear boots to gain contact. A high top load requires a tighter rear cinch than a snug side load. A fast gallop requires tighter cinches than a walk.

b. The exact amount of cinch pressure required can be determined only by experience. A safe rule to follow is to cinch with only sufficient pressure to hold the saddle and load in place. The fingers of the hand should pass easily between the cinches and the animal's belly. In testing cinch pressure the fingers should be inserted from rear to front, so that when they are withdrawn the hair will not be ruffed so as to cause a gall.

c. Final adjustment of cinch pressure is made after the load is placed upon the saddle and again at the first halt.

199. Adjustment of breeching and breast collar.—*a*. With the saddle cinched in its proper position, place the breeching upon the animal's croup so that the croup piece will be flat a few inches above the dock. Attach the two "hold-up" straps to the top D-rings on the rear edges of the saddle pads and adjust so as to hold the breeching in place. Adjust the two "lead-up" straps in rear to hold the breeching body piece horizontal. The two lower or "holding" straps should now be attached to the two bottom D-rings on the rear edges of the saddle pads. The holding straps should be adjusted to admit an easy and free swinging motion of the breeching body around the buttocks at each gait without undue pressure or binding. Both straps must be adjusted to the same length. The animal should be moved about at the different gaits to assure a correct adjustment.

b. The breeching functions through the holding straps by transmitting pull both to the croup piece and the breeching body. The holding straps should be adjusted tighter than the top or hold-up straps. The only function of the hold-up straps is to hold the breeching up and in place, and if these straps are too tight the breeching cannot function and will be pulled up out of place.

c. The breast collar for cavalry pack transport should be adjusted so that it is snug when the horse extends his forelegs at the gallop.

200. Unsaddling.—*a*. If the free end of the running strap has been passed through the metal loop, to uncinch remove it. Release the running strap by pressing downward on the metal loop, keeping the hand close to the animal's belly; with one hand on the lower bar of the cinching device, pull outward which releases the straps, then the device may be unhooked. The cinches being unhooked, they may be thrown over the saddle from the right side. Move the saddle to the rear a few inches to free the breeching; then raise the breeching over the animal's croup and place it upon the saddle.

b. The saddle is removed by a packer on each side taking hold of the foot rests and lifting the saddle over the croup of the animal. The mohair pad is then removed, folded with the wet side in, and placed on the saddle.

201. Adjustments of saddle pads.—No adjustments of the saddle pads are required for pack animals of normal conformation and good condition. Some adjustments, however, are necessary to secure and maintain proper fit on animals of poor conformation or those that have lost flesh in campaign, and also to relieve injuries.

a. To prevent injuries.—(1) Unless the animals being fitted are clearly of abnormal type for pack saddlery, they should be saddled and worked to determine the bearing of the pads while breaking-in the saddles. The best time to check the bearing of the pads is immediately after unsaddling when sweat marks are present. Excessive bearing is indicated by heavy sweat marks with the hair of the animal either ruffled or compressed, depending on the movement of the saddle; dry surfaces on the animal indicate little or no bearing. The saddle pads should be adjusted for uniform bearing, generally by first relieving the pressure caused by excessive bearing. The animal should again be saddled and worked so as to check the adjustment definitely. When excessive bearing has been reduced, the area of little or no bearing should be built up to give uniform bearing.

(2) Abnormal types of pack animals should be saddled without the blanket or woven saddle pad and the fit observed from the front, sides, and rear. The front of the saddle should fit smoothly against the animal with no compression of the withers. The sides of the saddle should not be pushed outward excessively. The rear of the saddle should follow its natural curve without pinching. There should be uniform bearing of the pads along the weight-bearing muscles of the back as viewed

from the front and rear. Excessive bearing or compression at any one place is especially to be avoided.

The excessively thin withered animal and the sway-backed animal require the addition of hair to build up the pads. The excessively thick, flat, broad-withered type requires the removal of hair. For the otherwise normal but short-backed animal, hair need be removed from the rear of the pads only. For the broad-ribbed and large-barreled types, hair need not be removed unless these features are excessively large. When the animal obviously should not be worked to obtain sweat marks, the method of determining an accurate fit consists simply in marking the animal so as to transpose the markings to the saddle pads. In cases requiring the addition of hair the animal should be marked on all sides of the area and then saddled, without blanket or mohair pad, the saddle cinched tightly and left in place for a few minutes so as to transpose the markings. The dry or unmarked spots inside of the marked area indicate where hair should be added. If in doubt just where to confine the marking, the entire position of the saddle should be marked. In cases requiring removal of hair, mark the spot or area of the animal and saddle as stated above. The markings on the saddle pads will indicate where hair should be removed from the pads. Water may be used for marking, though as a rule it does not give sufficiently accurate marks on the felt pads. Chalk, colored grease, white zinc oxide, or methylene blue give more distinct markings.

(3) Adjustments for maintaining the fit for animals that have lost flesh in campaign consist merely in replacing with hair in the pads the flesh lost by the animals, to conserve the original bearing of the saddle pads.

b. To relieve injuries.—(1) With properly fitting saddlery and hanger positioned loads there is but small chance for either the saddle or the load to be a primary cause of injury. Many injuries are relieved by correcting a faulty position of the saddle, by readjusting the breeching or breast collar, and by attention to cinch pressure and loads. A minor bruise would not ordinarily require the removal of hair from the pads, nor would removal of hair be necessary for a swelling that has subsided, unless the swelling was caused by a lump of hair in the pad. It is often possible to relieve pressure by tightening a thong instead of removing hair.

(2) Basically all pack saddle injuries may be relieved by reducing or removing pressure or friction from over the injured part. This may be accomplished by chambering or making a recess in that part of the pad immediately over the injury.

(3) To chamber a pad, detach it from the frame and loosen all thongs passing through the marked area so as to have a chamber of at least one inch greater radius than the marked area. Tie loose knots in the ends of the thongs. Pull the leather slip out of the handhole nearest the marked area. Use the hair hook in the tool kit and pull the required amount of hair, pulling small amounts each time from the back of the pad. Press in the chamber on the contact or felt side, and tie the thongs. Have an assistant push the thongs inward with the handle of the hammer when tying them. Check the accuracy of the work by holding the pad upon the animal in the exact position it would occupy, observe the chamber, and see that all pressure would be relieved when the pad is attached to the frame. Repeat this checking with the pad attached to the frame.

(4) As soon as the injury is healed the normal shape of the pad should be gradually restored. Any suitable material may be used if curled hair is not available. The resiliency of the curled-hair pads may be restored by pounding the pads with the clenched fist.

c. To make adjustments.—(1) The pads must be detached from the frame before adjustments are made. They are held to the frame at the top by two bronze staples that are engaged to two staple hooks of the frame, and at the bottom by the bottom bar pockets. To detach a pad, press downward on the head of each bottom bar pocket pin, turning the pin to the left until the lug on the pin springs upward through the keyway of the pocket, and remove the pin. When both pins are removed, take hold of the cinch ring with one hand, and with the other hand near the edge of the pad push the lower part of the pad from the frame. Swing the pad to the middle of the frame and remove it by lifting it off the staple hooks at the top of the frame.

(2) To attach a pad, turn the saddle frame on its back, put the canvas saddle cover in place, leaving room in the openings of the cover for attaching the staples of the pads to the staple hooks of the frame. Attach a pad by holding it near a line through the middle of the saddle arches, away from the bottom bars of the frame, and engaging the staples to the hooks, lifting upward to secure them. Hold the pad engaged and bring it against the bottom bar of the frame. In case one man is attaching the pads, rest that side of the frame with the attached pad on the ground. Repeat for the opposite pad. Hold the pads to the frame and turn the saddle right side up, then straighten the cover by swinging the pads toward the middle of the frame and pulling the cover downward. Swing the pads back and engage them to the bottom bars of the frame. The bottom bars are pressed into the pockets of the pads by holding one hand on the foot of each pocket in turn and pushing the pad outward toward the bottom bar with the other hand. Insert the bottom bar pocket pins in the holes of the pockets and engage them by pressing down the heads and turning the lug on the pins one-quarter turn to the right, when the lugs should snap into place.

(3) Practically all adjustments are made through the handholes in the leather backs of the pads, thus conserving the smooth contact surface. The amount of hair to be added or removed may conveniently be determined by holding the pad in position on the animal where the fit may be observed. The pad must be held in the exact place where it will bear when it is attached to the frame.

(4) To determine the exact place on the animal's back on which to hold the detached pad, make a vertical chalk mark on the animal along the front edges of the pads when attached to the frame; also a horizontal mark to indicate the lower edges of the pads.

(5) Remove the leather slip in the handhole nearest the place where the adjustment is to be made and pull the thong out. As the pads are formed and tied in a mold, it is important to untie only the minimum number of thongs. Thongs may be removed if desired, in which case it will be necessary to use the awl of the saddle kit to replace them. Thongs may be left in place by tying loose knots in the ends. If

hair is to be added through a handhole, use the 12-inch stuffing rod of the saddle kit for pushing the hair to place; if hair is to be removed, use the hair hook. Hair should be well loosened before stuffing and only small amounts put in at a time. It should be firmly compressed into place, being careful not to push the tool through the felt contact surface. The handle of the hammer is handy when pushing large amounts of hair through a handhole.

(6) For minor adjustments hair may be pushed through a tuft hole on the contact side by first removing the thong. The short 8¾-inch stuffing rod is used for this work. Only small amounts of hair at one time may be pushed through the tuft holes, and care must be exercised not to tear the felt.

(7) In removing hair, pull small amounts each time from near the leather back of the pad and push the contact side inward to form the required chamber or contour. Thongs must be tied tightly after making the adjustment. To tie a thong, have an assistant push against the thong on the contact side with the handle of the hammer when tying it, or rest the thong on a stick or rock.

(8) After adjusting the detached pads, they should be attached to the frame and the accuracy of the adjustments checked by again saddling without the mohair pad. In fitting the withers, space must be left for the mohair pad.

Section IV

LOADS AND LOADING

202. Pack loads.—The Phillips saddle carries either hanger-positioned or rope-hitched loads.

a. Hanger loads.—(1) Side hangers are equipped with hooks which fit over the hanger bars of the saddle. The hangers rest on the spring steel ribs and are tied down to the two end staples on the bottom bars. Hooks for hangers should fit loosely. All hangers, boxes, etc., should be equipped with distance pieces or rests riveted to the lower part of the box or hanger. The height of these pieces should be sufficient to clear the load from hooks to distance pieces. The load should never rest on the swell of the ribs. The "hanger distance," or the measurements between the outside edges of the hooks, is 13⅝ inches for all Phillips pack saddles, cargo or cavalry.

(2) In designing hangers for side loads of unequal weight, the heavy and bulky loads should be placed where they will ride best, then the lighter and less bulky loads fitted so as to obtain the desired balance. Where there is but a slight difference in the weight and bulk of side loads, the heavier of the two loads will balance slightly above the lighter load. But where a bulky side load extends outward considerably from the side of the saddle, although somewhat lighter than the opposite load, it will generally balance higher on the saddle. Correct balance should be maintained at all times. Heavy contents of boxes should be placed in forward compartments of the boxes. All loads should be close to the saddle and not extend downward much below the middle of the saddle if practicable.

(3) The arches of the saddle frame are designed to admit the attachment of top load arches. There are two standard types of load arches; one is for all heavy loads, such as the loads of pack artillery, etc., and the other is for light loads. Both types are bolted to the saddle through the two holes in each saddle arch. High top loads should be avoided if possible. A slight excess of the weight of loads should be placed just forward of center. Special types of load arches are occasionally required for special loads.

b. Rope-hitched loads.—Any form or type of hitch may be used on this saddle. The foot rests on the lower corners of the pads and the hitch hooks on the bottom bars simplify the forms of hitches.

203. To lair up packs.—Spread a manta evenly on the ground and place the bundle diagonally across the center. Standing on one side of the bundle, pick up the manta by the nearest corner, bring it snugly over the bundle, and place one or both knees on top; reach over, grasp and fold in the opposite corner, then bring the double portion over the top of the bundle so that the edge will come nearly across. Place the left knee on top of the bundle to hold the manta in position, crimp the manta in at one end and bring the flap over the bundle. The operation is repeated for the other end, folding it under so that the folded edge will come near the middle and on top.

Grasp a lair rope (⅜-inch rope, 30 feet long, at one end of which an eye has been spliced), and form a loop by passing the end through the eye, then place the loop lengthwise around the middle of the bundle and draw taut so that the eye comes near the top of one end. Take a half hitch around the bundle near one end, one in the middle of the bundle, and one around the other end, after which the remainder of the rope is carried around the bundle lengthwise and tied on top by two half hitches. Packers work singly in lairing mantas on cargo.

204. To sling and double-tie the load.—*a.* Two packers work together. The left packer takes a sling rope and doubles it near the center, passing the ends over the saddle to the right side, letting the loop hang down on the left side. He then spreads the rope apart to about 3 inches from each edge of the saddle. (See fig. 22.)

b. The right packer then places his bundle well up on the saddle, resting it on the sling rope, a flat side of the bundle being in contact with the saddle, and holds the bundle with his left hand. The left packer then picks up his bundle and places it on the saddle so as to overlap about 4 inches on top of the other bundle and holds with his right hand. The right packer then passes the ends of the rope over the top of the load. The left packer receives them, and passing the rear rope through the loop, takes out the slack and ties a square knot. He then similarly ties a square knot with the forward rope. (See fig. 22.)

c. The load is then settled in place as follows: the right packer holds his bundle in position while the left packer lifts upward and outward on the bottom of his bundle so that the tops will come even, after which he lets go and each packer grasps the top of his bundle and settles the load by pulling down upon it. The left packer coils the loose rope ends and the loop and tucks them under the sling rope on top of the load. The load is now ready to lash. The method of settling here described is for evenly balanced side loads. A heavier load on a side overlaps a lighter load for the required transverse balance. All loads should be positioned so as to have a slight excess of weight forward.

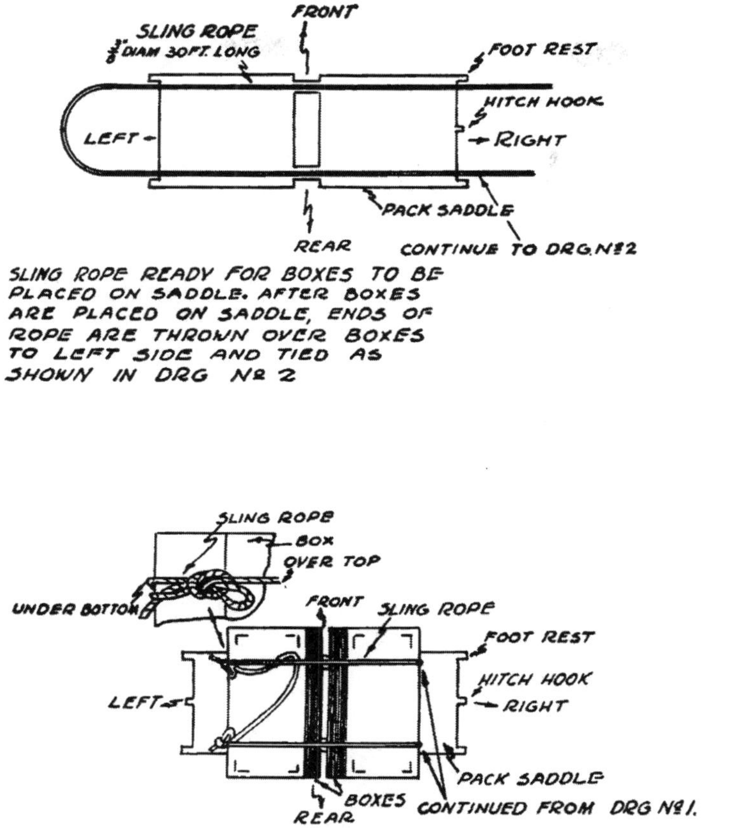

FIGURE 22.—To sling and tie a load.

205. Adjustable lash rope.—This lash was designed for use with the Sweeten diamond hitch or for use as a load cinch. It consists of two 4-inch rings, to each of which are spliced 25 feet of half-inch manila rope. The rings are connected by 9 feet of half-inch rope, one end of which is spliced to one ring, the other end being connected to the same or opposite ring by a metal snap. For ordinary loads the rope with the snap is passed through the opposite ring and snapped to the first ring, giving a load cinch approximately 4½ feet long. For very small loads the rope with the snap is passed through both rings and snapped to the opposite ring, thus giving a load cinch approximately 3 feet long. (See fig. 23.)

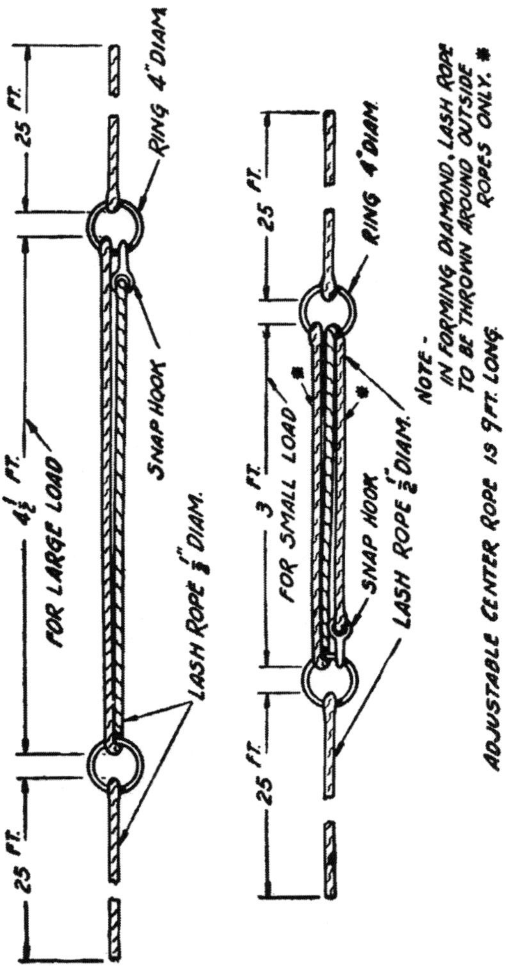

FIGURE 23.—Adjustable lash rope, Phillips pack saddle.

206. Diamond hitch.—For the diamond hitch, or modified forms of it, the lash rope has a loop in one end; this loop hooks to a hitch hook on the bottom bar of the saddle, the diamond then being formed in the usual manner, the lash rope passing under the hook on the opposite side of the saddle and under the foot rests on the lower corners of the pads. This hitch was developed for use with aparejos. It has a tendency to pull the sides of a saddle outward.

207. Sweeten, Nagle, and Phillips cargo hitches.—The design of the Phillips pack saddle has enabled the development of some new and superior types of hitches.

a. Sweeten diamond hitch.—The Sweeten diamond hitch is of special value for pack trains. It requires a special lash rope, but this rope is of particular value when a load cinch is required, as for all pack artillery loads.

(1) *First step* (fig. 24).—The lash rope is placed over the center of the load with the 4-inch rings hanging just below each side load. The left side packer passes his lash rope

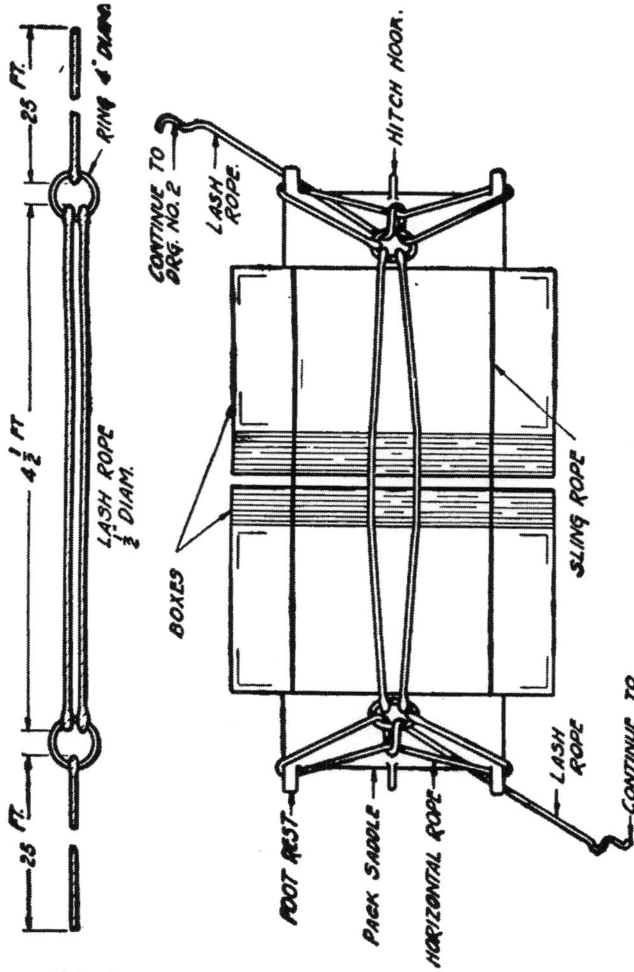

FIGURE 24.—Sweeten diamond hitch, Phillips pack saddle, No. 1 (first step).

over and under the foot rest on the rear corner of the saddle, then under the front foot rest and through the 4-inch ring from outside to inside, then down and under the horizontal rope coming up on the outside, again through the 4-inch ring from outside to inside and under the rear foot rest. The right packer executes the same movement and, working toward his right, passes his lash rope under the front foot rest, then under the rear foot rest, etc., completing the first step with the rope under the front foot rest.

(2) *Second step* (fig. 25).—The left packer, with the lash under the rear foot rest, passes the lash over the rear corner

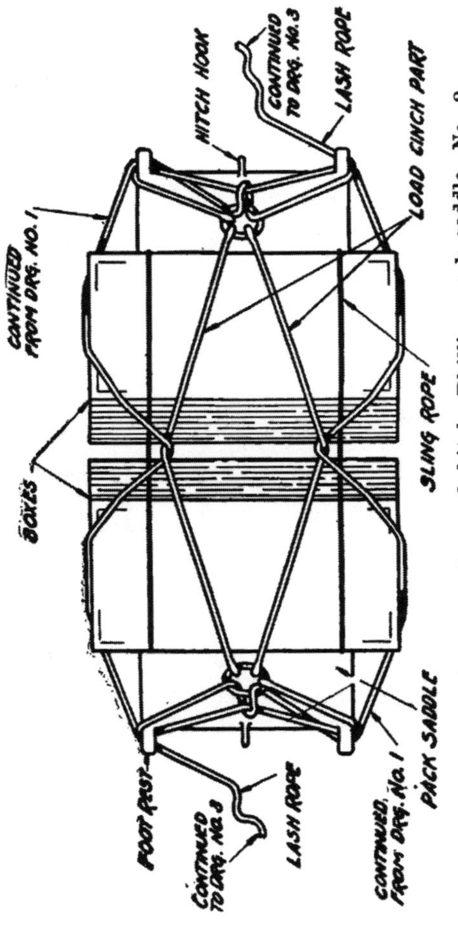

FIGURE 25.—Sweeten diamond hitch, Phillips pack saddle, No. 2 (second step).

of the load, then around the rear rope of the load cinch part, over and under to the packer on the right side, who places the lash over the rear corner of the load on his side. The right packer working toward his right performs similar movements. With the lash under the front foot rest he passes it over the front corner of the load, then around the front rope of the load cinch part, over and under to the packer on the left side, who places the lash over the front corner of the load on his side. This forms the diamond.

(3) *Third step* (fig. 26).—To tighten and tie the hitch, both packers take hold of the first rope passed through the 4-inch

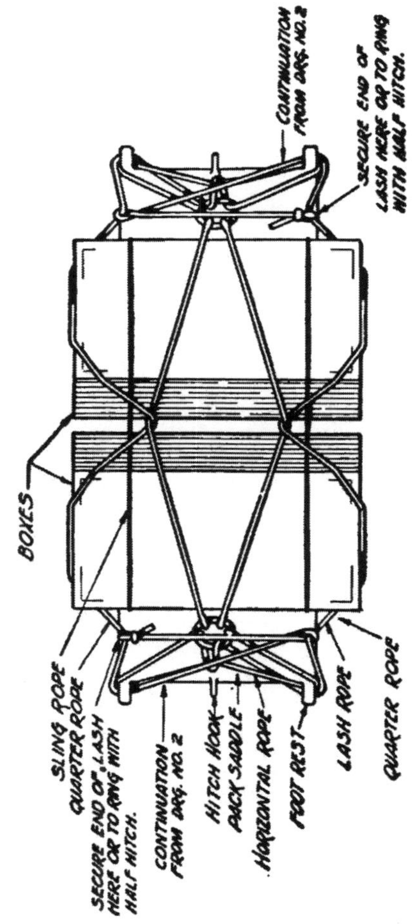

FIGURE 26.—Sweeten diamond hitch, Phillips pack saddle, No. 3 (complete).

rings below the rings and, calling "down," pull straight downward, taking up all slack. Holding the slack, each packer places his left foot against the bottom bar of the saddle to steady the saddle, and then grasping the lash with the right hand below the horizontal or standing rope, calls "Up." Both packers pull upward taking up all additional slack and bringing the standing rope toward the 4-inch ring. The packers hold the slack. The right packer, holding the slack of his lash rope in his right hand, takes one step toward the rear and grasping the lash rope from the left side with his left hand takes the slack from this rope and thus frees the left packer. Then, taking a step forward, he places his own lash (which is in his right hand) under the front foot rest (if it is not already there), and calls "Take." The left packer takes the slack in this the front lash of the diamond and, passing the lash under the front foot rest and then around the rear quarter rope from outside to inside, coming out under the lash, then around the front quarter rope, holds until the right packer has completed similar movements. The right packer, holding the slack of the rear lash of the diamond, passes the lash under the rear foot rest and around the front quarter rope from outside to inside, coming out under the lash, then around the rear quarter rope, when he calls "Tie." Both packers then place their left feet against the bottom bars of the saddle and take up the slack by drawing the quarter ropes inward and tie. (This braces, on a diagonal, the opposite packer, both together bracing cargo and animal.) The right packer ties his lash at the rear quarter rope, and the left packer ties his lash at the front quarter rope. A half hitch or other type of hitch may be used.

(4) *Variation.*—A variation of this hitch, known as the single hitch, is extremely simple and for certain types of loads very effective. It is necessary in such cases, however, to double tie the sling rope. The principal use for this hitch is as a load cinch. The hitch is practically the first step of the Sweeten diamond hitch, except that the lash ropes for the last stage of the first step are thrown over to opposite sides of the saddle instead of under the foot rests. The tie is made by passing the ends of the lash ropes around the bottom ropes under the rings to the hitch hooks or around the ropes above the rings. (See fig. 27.)

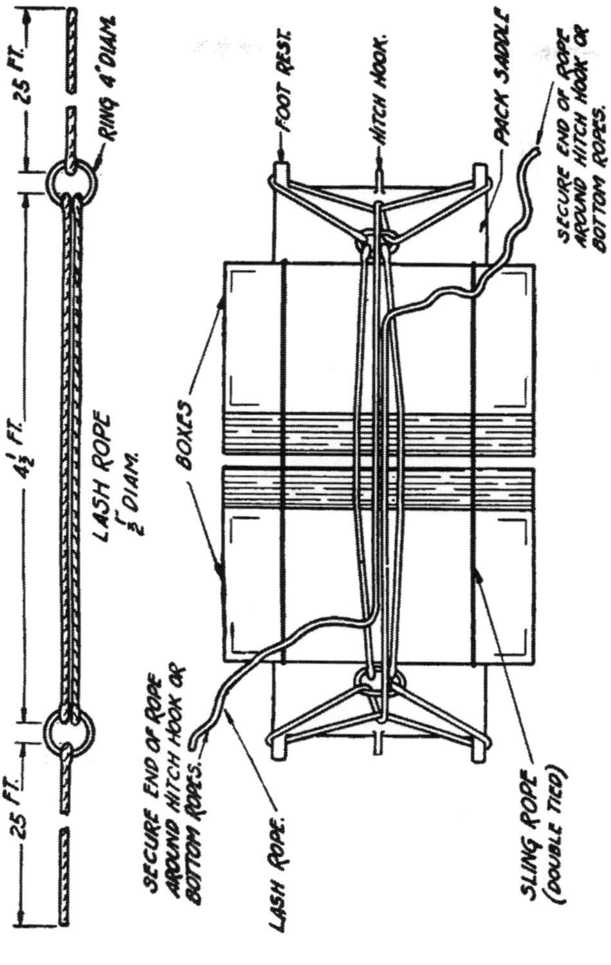

FIGURE 27.—Single hitch, Phillips pack saddle.

b. Nagle hitch.—The Nagle hitch is a modification of the Sweeten diamond hitch. It is a satisfactory hitch for two side loads and can be thrown more quickly than either the Phillips or Sweeten hitches. The lash rope for this hitch consists of two 28-foot pieces of ½-inch rope fastened to a 4-inch ring. (See fig. 28.)

(1) *First step.*—The ring is placed in the center of the load, one bight and its corresponding free end falling to each side. (See fig. 29.)

(2) *Second step.*—Each bight is now placed under the foot rests of the saddle on its side, passing over its free end between the two foot rests. The "ring-to-foot-rests" sections of each bight are now thrown out and around the corners of the loads. A firm tension is now applied downward to the free end. (See fig. 30.)

(3) *Third step.*—Next the free ends are thrown up and over the loads. Additional tension is now applied by the two packers, working together, one pulling outward just above the foot-rest-to-foot-rest position of the bight, the other taking up the slack as it accrues. (See fig. 31.)

(4) *Fourth step.*—The free ends are now made fast by any of the usual hog-ties, the first binding point to be between the vertical ropes and cargo. (See fig. 32.)

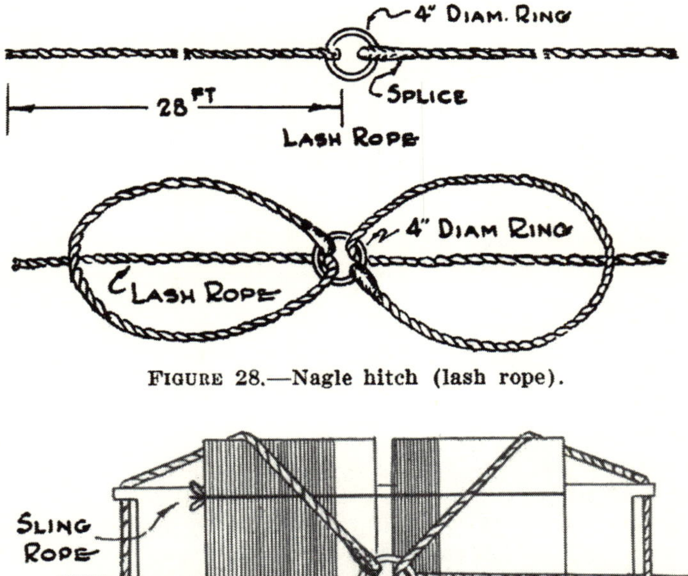

FIGURE 28.—Nagle hitch (lash rope).

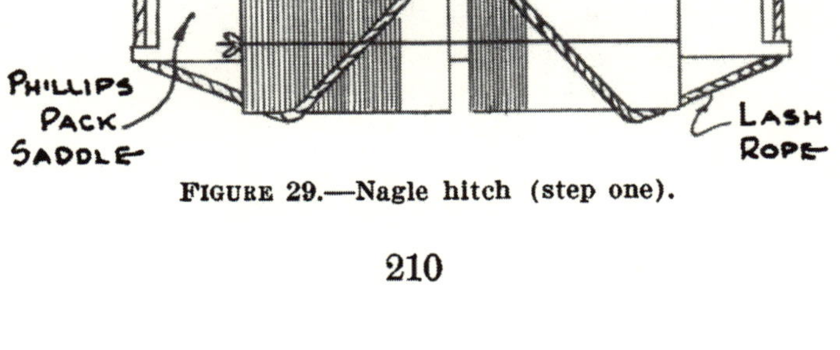

FIGURE 29.—Nagle hitch (step one).

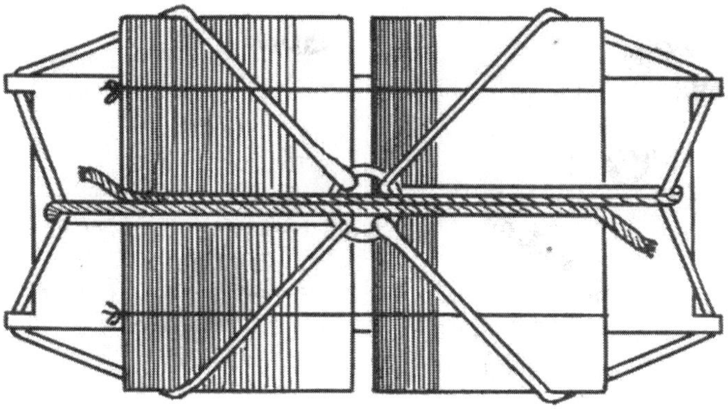

FIGURE 30.—Nagle hitch (step two).

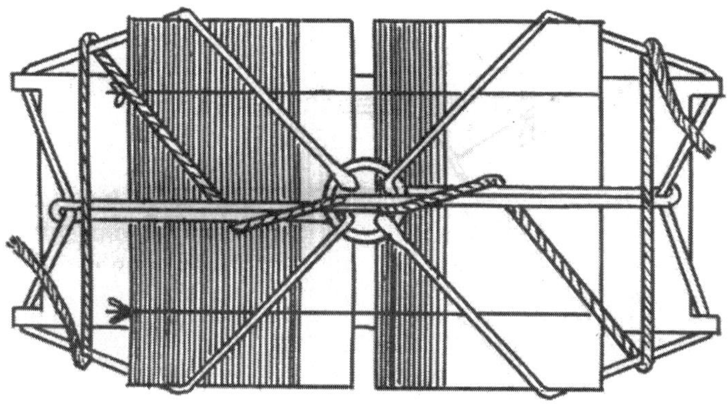

FIGURE 31.—Nagle hitch (step three).

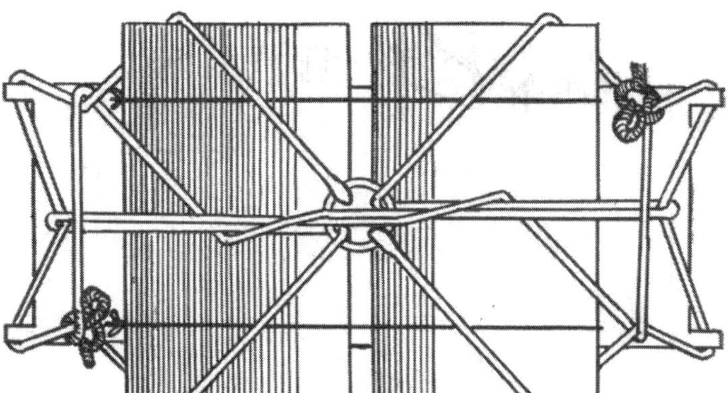

FIGURE 32.—Nagle hitch (complete).

c. Phillips cargo hitch.—The Phillips cargo hitch is for all arms and services, but is of special value for Cavalry, Infantry, and other arms. The lash consists of a 60-foot, ½-inch manila rope with a loop in one end.

(1) *First step* (fig. 33).—With the lash on the left side of the saddle, the loop is placed on the hitch hook and part of the lash is passed over the load to the packer on the right side, who places it under the hook from rear to front. The left packer then places the lash under the hook on his side,

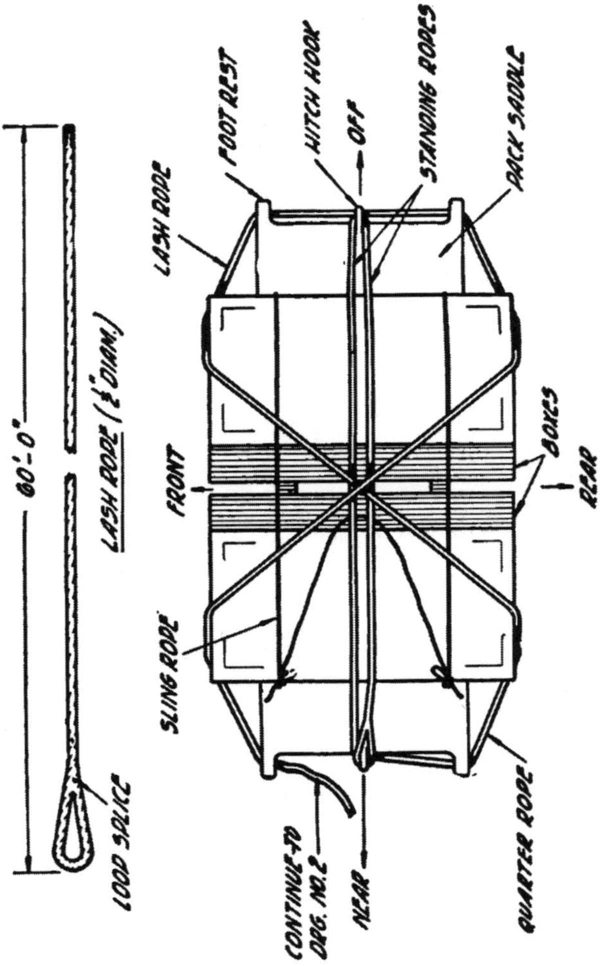

FIGURE 33.—Phillips cargo hitch, No. 1 (first step).

from the front, then under the rear foot rest and over the rear corner of the load diagonally toward the right front corner. The right packer, taking sufficient lash, places the diagonal over the right front corner of the load, under the front foot rest, then under the rear foot rest and diagonally over the rear corner of the load toward the left front corner. All slack is now taken up as follows: The right packer places his left foot against the bottom bar of the saddle and pulls the slack from the rear standing rope, the left packer taking the slack by pulling the front standing rope. The left packer then pulls the slack from the rear quarter rope, the right packer taking it from the front and then from the rear corners. The left packer takes this slack over the front corner of the load and places the lash under the front foot rest and holds with his left hand.

(2) *Second step* (fig. 34).—The left packer, with his right hand, back of hand outward, loops the lash around the rear quarter rope from outside to inside, coming out under the lash and leaving the end of the lash rope hanging outside the footrest.

(3) *Third step* (fig. 35).—The loop is grasped in the left hand and moved forward outside the standing ropes, then around the front quarter rope from the inside.

(4) *Fourth step* (fig. 36).—The end of the lash is placed through the formed loop from the inside and drawn tight, the loop remaining in front of the standing ropes. Holding this slack with his left hand, the left packer steps to the rear of the saddle, places his right foot against the rear footrest and pulls back. This takes up the slack. He then passes the loose end over to the right side in front of the standing ropes. The right packer takes all slack and then binds it by pulling the lash under the standing ropes toward the rear. He then grasps the lash with his left hand and forms a loop around the rear quarter rope from inside to outside, which he holds. With his right hand he pulls the end of the lash from inside the standing ropes and passes it around the front quarter rope from outside to inside, coming out under the lash, outside the standing ropes, then through the loop held by the left hand, when all slack is taken up by drawing the quarter ropes inward. The lash is tied with a half hitch or other convenient type of hitch. As some loads settle, due to stretching of the ropes

or settling of the cargo, it will generally be necessary only to untie the lash and tighten the quarter ropes to secure the load again.

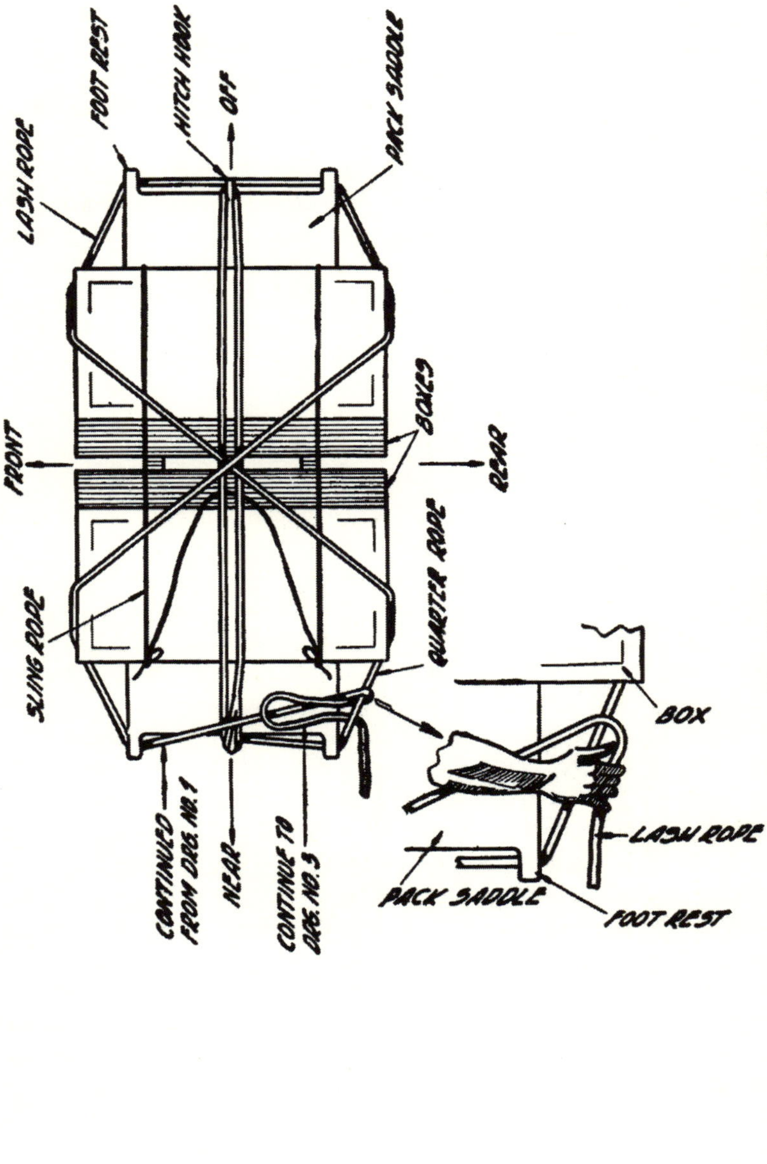

FIGURE 34.—Phillips cargo hitch, No. 2 (second step).

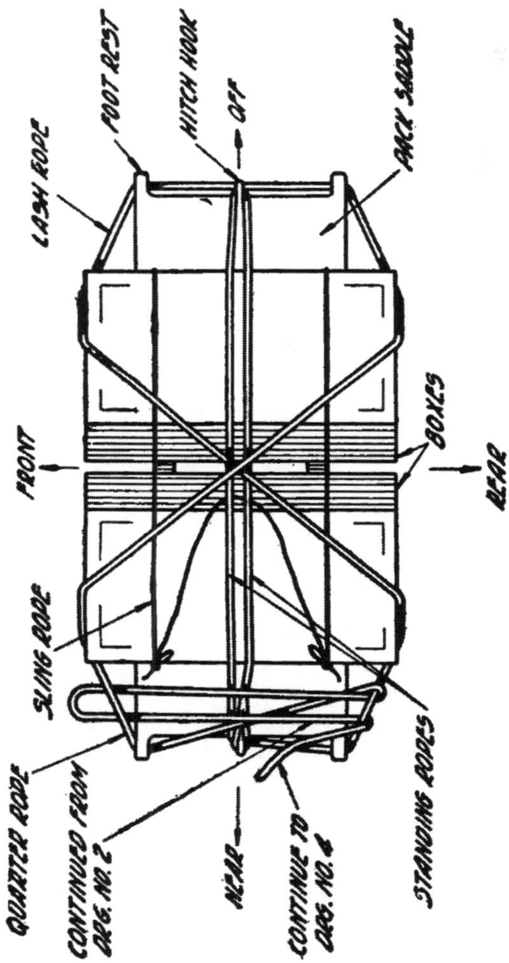

FIG. 35.—Phillips cargo hitch, No. 3 (third step).

208. Adjustment of loads and cinches on the march.—*a*. Rope-hitched loads settle rapidly shortly after commencing a march, thus loosening the pressure of the lash rope. The weight of the load causes the saddle to settle and this, combined with drawing up of the animal's belly, causes a looseness of cinch pressure. Cargo pack units should halt within the first mile, tighten cinches, and readjust the pressure of the lash ropes of all loads requiring it. Cinches should be tightened first.

b. Final adjustment of cinch pressure is made after the load is placed on the saddle, not when first saddling. Correct cinch pressure cannot be determined when the animal is first saddled.

c. Pack units equipped with hangers should adjust cinches at the first halt. When not practicable to remove loads at halts, great relief will be afforded the animal and injuries possibly prevented by releasing cinch pressure. The position of the saddle and cinch pressure should be inspected at each halt.

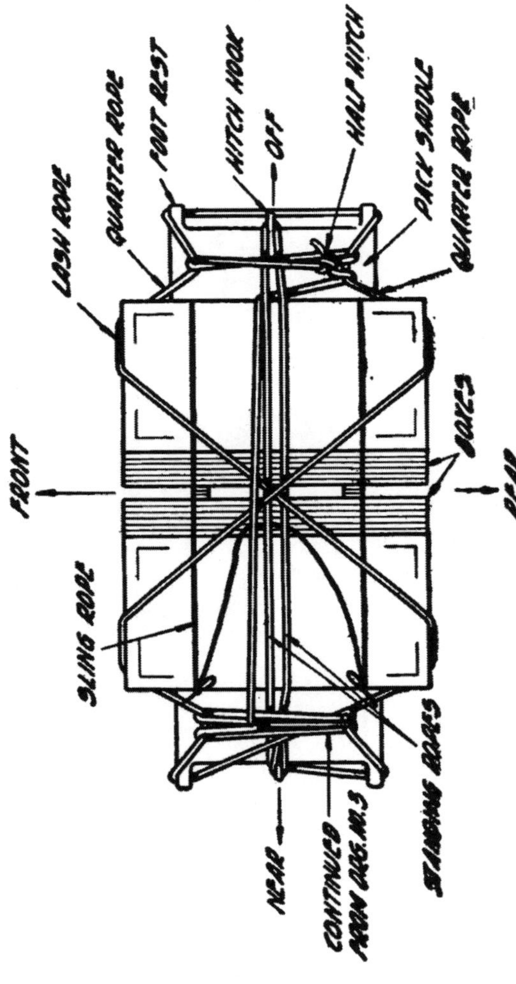

FIGURE 36.—Phillips cargo hitch, No. 4 (complete).

INDEX

INDEX

INDEX

INDEX

134205°—39——15

INDEX

○

Made in the USA
Middletown, DE
26 January 2023

23017608R00139